MODER
CHEMICAL MAGIC

by

JOHN D. LIPPY, JR.

and

EDWARD L. PALDER

This volume is the most complete and authoritative book that has been compiled on the subject to date. It has been several years in research and preparation. The co-authors are thoroughly versed in the art of magic and the science of chemistry. By combining the two old professions they have produced this new volume, MODERN CHEMICAL MAGIC.

For amusement in your own home or entertaining a group of friends, this new overture to magic has no peer; it is startling, mystifying and most effective. It gives a new approach to the professional or amateur magician to add variety to his programs. It serves the science teacher and scholar with something that shows the more interesting and dramatic side of the age-old science of chemistry.

Tricks, effects and methods are well explained. The chemicals are easily procured, the majority of them being in powder form, which makes them easy to handle and carry. MODERN CHEMICAL MAGIC is divided into fifteen chapters with a brilliant introduction by the world famous magician, Harry Blackstone. Contained herein are over two hundred and fifty tricks, methods and effects —from chemical daubs to chemical demons— truly chemical-magic miracles, all explained in detail and in some instances illustrated.

Concerning the Authors..

John D. Lippy, Jr., known for many years professionally as *Lippy, The Magician,* but today appearing occasionally with his charming wife as "The Lippys, Amazing Master Magicians," has been associated with the art of necromancy for over fifty years. He numbers among his friends all of the greats in magic, had a large troupe on the road in the early twenties, and appeared at the White House before the President. He has also entertained many national celebrities.

John D. Lippy, Jr.

Mr. Lippy is a chemist by profession—studying at Gettysburg College under Dr. Edward S. Breidenbaugh, a noted teacher and authority on chemistry. He has written considerable on the science as well as being the inventor of many chemical toys and novelties.

Palder's interest in magic began some twenty-five years ago, but for the first decade of this thriving interest it remained for something highly dramatic to change it from a passive desire to "do magic" to an active participation. It was when he decided to bring in a magician to help defray the lagging costs of the school newspaper that he had his tie "cut and restored" by this man of mystery. A few years later he found himself in many foreign lands at the "invitation of Uncle Sam" and he took the opportunity to learn of the magic of other countries.

Edward L. Palder

No claim is made of ever seeing the famous Indian Rope Trick, but the magic of other countries developed the incentive to conduct shows on a semi-professional level.

With his graduation as a Chemist and Pharmacist, the idea was born to write a book using the basis of Chemistry as the "secret element" for doing magic. You will enjoy the material contained within these pages as much as he has in preparing it.

MODERN CHEMICAL MAGIC

by

JOHN D. LIPPY, JR.

and

EDWARD L. PALDER

Dedicated
To All
Those
Magicians
Who Have Made
Magic
An Art

THE SUBJECT MATTER of this work is based on the scientific. The authors are specific in directions, explanations and warnings. If what is written is disregarded, failure or complication may occur—just as is the case in the laboratory. In the instances where flammable or toxic chemicals are involved, explicit explanation and warnings are provided.

INTRODUCTION

by

HARRY BLACKSTONE

World Famous Magician

The alchemists of old searched endlessly for ways to turn base metals into gold. They cudgeled their brains for solutions to such problems as the finding of a universal solvent—they treated the occult as a cozy, friendly, and helpful reality.

Back to the dawn of history stretches the preoccupation of man with magic. In biblical times Pharaoh commanded magicians to solve puzzling dreams for him—dreams that a young Hebrew named Joseph finally translated into the first ever-normal granary on record.

Now, twenty centuries after the birth of Christ, mankind is slowly learning to separate the arcane from the mundane. Now with the birth of space vehicles, we are learning that the problems of continuous never ending motion are indeed soluble—not by the black arts, but by scientific application of physical law.

In the twentieth century, man no longer needs to make a Faustian pact with the Devil to involve worldly powers. Atomic fusion is not magic, it is physical. Faust and Mephisto remain—but, far from being characters in a medieval horror story, Mephisto turns out to be the genial Mr. Applegate, and Faust, a middle aged man who dreams of beating the Yankees.

Indeed now, we hear that a foreign physicist has solved the age old problem of the alchemists—and the claims to have transmuted lead into gold. And yet magic stays with us—not black magic, but a magic that entertains—magic that instructs—magic that, at its best, thrills.

Our world is a chemical world built on physical processes, far from the world existing in the minds of the alchemists. And yet the magicians—no less than the laboratory technician—can unlock the secrets of chemistry.

The black magician of the Dark-Ages with his newts and skulls and black masses is as far removed from chemical magic as a cooking fire in a cave at the dawn of history is from a nuclear reactor. And as close, too, because there is a relationship.

Modern magicians and modern scientists—have discovered what the alchemist of old could never learn—the simple fact that natural laws are more remarkable and more productive of results than supernatural laws could ever be.

But chemical magic owes much to the alchemist. For with all his fantastic rigamarole, with all the superstitious, even blasphemous-ritual, the alchemist was the forerunner of the modern orderly study of chemistry as we know it today. The best monument to this fact is modern chemical magic, dramatizing as nothing else could possibly do, the age old science of chemistry.

In one enlightened age, we welcome that which surprises, that which awes, that which mystifies. Trolls, gnomes, and ogres no longer awe us. Grimalkin has become the black cat which we amusedly let cross our paths. The alchemist has come out of his garret, and now regales us with surprising, awesome, and mysterious phenomena which come, not from the spirit world, but from an even more amazing place—the world around us.

Indeed Alchemy had two heirs. Their names are Chemistry and Magic. And they serve us as surely as the Djinn served Aladdin.

Harry Blackstone

FOREWORD

Modern Chemical Magic is a very fine reference for ambitious teachers who want to popularize scientific information by presenting dramatic experiments. This book should be useful for science teachers.

CHARLES H. BOEHM
Superintendent of Public Instruction,
Commonwealth of Pennsylvania,
Harrisburg, Pa.

The first definite work on a major subject so closely associated with chemistry and magic in over twenty-five years. I fully recommend it for the serious student in the fields of chemistry and magic and offer my congratulations to the authors for preparing this book.

DR. JOHN E. KIEFFER
President,
Universal Research and Consultants, Inc.,
Washington, D. C.

Please accept my congratulations on the thoughtful study that must have gone into the planning and writing of *Modern Chemical Magic*. It provides an exciting introduction to the wonderful world of entertaining with science.

WILLIAM W. SAUNDERS
Chemistry Department,
Montgomery Blair Senior High School,
Takoma Park, Maryland

Modern Chemical Magic . . . Teachers often make chemistry a routine course, filled with drill and drudgery. It is very refreshing to find such a broad collection of chemical experiments that show how the study of science can be fun. The reader will be tempted to try these experiments. They are safe experiments worth trying.

KEITH C. JOHNSON
Supervising Director,
Department of Science,
Public Schools of the District of Columbia

I have examined the manuscript for your new book, *Modern Chemical Magic*. It is my opinion that this work can be particularly useful to teachers who are seeking ways and means for arousing the curiosities of pupils to study science and particularly chemistry.

The book is timely in that at the present time our country needs many chemists. Any item that serves to motivate young people toward further study of chemistry is therefore helpful. I regard this book as such an item.

Modern Chemical Magic can also be useful to chemistry-club members who wish to prepare a program that is both interesting and suggestive of chemical reactions.

Individual pupils who have some chemistry backgrounds and are careful with the use of chemicals will find the book helpful, too.

This book could be used even by experienced teachers to provide motivating devices for introducing new units, ideas for assembly programs or PTA meetings, a change of pace in club programs, or to work up an attention-getting lecture-demonstration for an open house, community or service club. With a little showmanship, a good public-relations program could be developed when the science department is called upon to entertain.

LEE E. BOYER, *Advisor,*
Science and Mathematics Programs,
Commonwealth of Pennsylvania,
Harrisburg, Pa.

ACKNOWLEDGMENTS

This book could not have been written without a great deal of assistance from a great many people, many more than we can hope to name here. It could not have been written at all had not the following made available to the authors some of their greatest effects developed in recent years. In this case this included nearly all of their most significant material in the field of "chemical magic."

To the following we again express our sincere thanks and gratitude:

HAROLD M. PORTER (CHEMCRAFT CHEMICAL CO.)
EARL C. LEAMING
L. L. IRELAND
BURLING "VOLTA" HULL
ORVILLE MEYER
LLOYD JONES
ROBERT SCHULMAN
S. W. REILLY
IRV WEINER
U. F. GRANT
CYNTHIA ZABREK

CREDITS

Harold M. Porter (Chemcraft Chemical Co.): A Magic Pitcher of Grenadine; A Magic Pitcher of Bluing; Water to Sherry Wine; Sherry Wine to Water; A Magic Pitcher of Wine and Bluing; Pouring Red, Yellow and Black from the Same Pitcher; Pouring Blue, Gold and Red from the Same Pitcher; Pouring Nine Different Colors from the Same Pitcher; Water to Milk and Milk to Ink; A Synthetic Cow; Conjuring Red from Blue; Water to Wine to Ink to Water; Pouring Ink and Milk from the Same Pitcher; Water to Milk, Water to Wine and Wine to Milk; Chemical Color Chase; Magic Blue Writing Paper; Writing That Disappears By Magic, and Salt Writing.

Earl C. Leaming: Character and Knowledge, Fugitive Color, Multicolor with Flash Finish, Burning Ice, Smoke Screen, The Hot Gin, The Fire Bowl, Self Lighting Candles, The Magic Soda Fountain, Magic Dyeing, The Chemical Garden, Clock Reaction—Yellow, Clock Reaction—White, The Backslider, The Money Maker, Spirit Writing.

L. L. Ireland (Yearbooks): Wine to Beer, Calcification of Handkerchiefs, Stage Size Out to Lunch Trick, Bottle of Smoke, A Self-Lighting Candle, The Escaping Shadow, A Flash Paper and Match Stunt.

S. W. Reilly: Famous German Spy Formula; Green Luminous Paint, Blue Luminous Paint, Violet Luminous Paint, Deep Yellow Luminous Paint; A Ghost Show Routine, Ghost Card, Ghost Coin; Mysterious Dice; Mystery Magazine; Mentally Speaking; Evil-Eye; Spooky Lights; Apparitions, Ghostly Faces, Ectoplasm, Seeing Ghosts.

Irv Weiner: Magic Methods with Daub.

Burling "Volta" Hull: The Scientific Secret, "The Spirits Know All," A Daring Volta Method, Dead or Alive Discovery, A Number Test, The "Spirits" Correct A Mistake, A Mind Reading Number Divination, Another No Flap Method, Spirits from China, Additional Ideas . . . One of the Three Little Pigs, The Rising Card, Novelty Balloon Stunt, Magic Hair Restorer, Other Ideas . . . Sensational New "Beer Powder," The Magic Soda Fountain, Magnet-Tizo.

Walter Essman: Patriotic Liquids and Silks.

Dr. Howard B. Kayton: The Bewitched Paper.

Teral Garrett: Dark Eyes.

Robert "Bob" Carver: Misdirection.

Orville Meyer: Think! Ink!

U. F. Grant: The Atomic Bomb, The Magic Pipe, The Pirate Knife.

Glen G. Gravatt: Gravatt's Mental Mystery.

Dr. Wm. M. Endlich: Ink and Water Columns.

Sid Fleischman: The Trick Without A Name.

PREFACE

Many books on magic have been written and many more are still to be written. Some of these are by authors recorded in the annals of history; others are by those seldom-heard-of authors, whose only result will be the tireless effort of preserving the antiquity of magic.

Be it a general book on all phases of the art, or an "encyclopedia," more correct to call them anthologies, they all have achieved a decided purpose.

From the magic of the past, from the masters of yesterday and today have come a legacy bequeathing the secrets of this ancient art to all.

In the years of research spent in the preparation of this book, it was difficult to find any one book, paper or manuscript devoted to "Chemical Magic." There were a few, however, limited in the scope of the material contained within their pages. There were also numerous copies of the originals.

Thus was born the idea for this anthology. A collection of magic centered about chemicals. No credit is claimed for the effects being original where credit is not given. Thus, where possible, credit is given to the original source of information.

Our only desire in preparing these pages is the knowledge that you, the reader, will help to preserve the antiquity of magic in the years to come.

In doing magic with chemicals, it often becomes an element of simplicity, for doing this type of magic, sleight of hand and complicated moves are at a minimum. What then is the necessary element for success?

This hidden element is known by many names, but for the present a distinguishable mode of approach is to simply call it "perfection."

The results of perfection are gratification. As a novice in the art of creating an illusion (not magic), he transcends from a period of beginning to what he might call "perfection."

So many misuse the word perfection. As applied to the art of legerdemain, it should have its own specialized meaning. Perhaps an adequate explanation might well be described as it being the most important part of any magician's repertoire in his ability to exhibit showmanship.

Contrary to general opinion, the "tricks" and illusions employed by magicians do not and will not determine success.

To illustrate this, let us consider the magician on the stage. Often times the illusionary effect is accomplished back stage by "mechanics" or on stage by assistants, yet it is the magician who receives the high acclaim of success by the audience.

One might rightly wonder why this is so. He might further wonder, when he as a magician has achieved what he thinks is perfection, he watches another magician perform the same illusion, he slowly realizes that his method of presentation and performance is not perfect.

When this occurs, one rightly wonders what is lacking. Is it speech, scenery, costumes, lighting or stage settings? Perhaps a summation might be that perfection arises from the use of one's natural ability not to do magic, but to successfully create an illusion.

Showmanship is only one element of success. In addition to the factors previously mentioned, undoubtedly the one most overlooked is simplicity.

Like the actor (all successful magicians are actors) who handed the spectator a card saying it was the ace of spades, the spectator without looking to verify that it was the ace accepted the statement because he believed the magician. A successful illusion was thus created by the use of simplicity and faith.

Perhaps one of our greatest magicians (long now since called forth by the greatest magician of all) made use of simplicity more than anyone.

The late Ted Anneman once made a statement: "To strive for simplicity, is to practice it. It requires but one factor—nerve."

The topic of discussion does not permit the illustration of the long-gone, but not forgotten, Anneman. It is safe to suggest, however, that for those magicians who strive for perfection by the use of simplicity read a few of his many books. They are a true heritage to the art of magic. They represent a legacy bequeathing a tradition of perfection to all.

As in the case of the author-magician just mentioned, there are others who have recorded and passed on to all their lifetimes of work midst a few pages of paper and ink. Consider and reflect upon the opportunities presented to you by Houdini, Thurston, Kellar, Leipzig, Blackstone and countless others. Take advantage of the opportunities present.

Never will one find so many golden opportunities for the attainment of success in the perfection of an art treasured so in the annals of the minds of men and revered so in their hearts.

The art of illusion is intended to deceive people, but the deception is intended to cause entertainment. People forget their troubles as they are being entertained. So do the job well, for ours is a responsible task—responsible to our past traditions, and responsible to the present and to the future for a job to be created successfully.

So as you strive for perfection in the creation of illusion, in order to successfully fool people, remember not to do tricks, do magic and enjoy the reputation that success brings forth.

JOHN D. LIPPY, JR.
EDWARD L. PALDER

CONTENTS

CHAPTER I. CHEMICAL MAGIC WITH LIQUIDS

Water to Wine 1
A Magic Pitcher of Grenadine 1
A Magic Pitcher of Bluing 2
Water to Sherry Wine 2
Sherry Wine to Water 2
A Magic Pitcher of Wine and Bluing 3
Pouring Red, White and Blue from the Same Pitcher 3
Pouring Red, Yellow and Black from the Same Pitcher 3
Pouring Blue, Gold and Red from the Same Pitcher 4
Pouring Nine Different Colors from the Same Pitcher 4
Water to Milk and Milk to Ink 5
Wine to Beer 5
A Chemical Clock 6
A Synthetic Cow 6
Conjuring Red from Blue 7
Water to Wine to Ink to Water 7
Pouring Ink and Milk from the Same Pitcher 7
Another Water to Wine 8
Character and Knowledge 8
Fugitive Color 9
Water to Milk 10
Water to Wine and Wine to Milk 10
A Magic Pitcher of Milk and Ink 10
Chemical Color Chase 11
Water to Wine 11
Multicolor with Flash Finish 11
Water to Beer 12
Rainbow Colors 12
Crème DeMenthé 13
By Command—Water to Ink 14
A Liquid from Solids 14
Chemical "Ice" 14
Chameleon Mineral 15
That Chameleon Mineral Again 15
Disappearing Whisky 15
Solid to Liquid 16

Blue to Red, Green, Crimson or Purple 16
Green to Blue 16
Blue to Red 16

CHAPTER II. FUN WITH DRY ICE

CHAPTER III. INVISIBLE INKS
Red 19
Blue 20
Black 21
Brown 21
Yellow 22
Green 22
Violet 23
Water Creates Visibility 23
Light Creates Visibility 23
Chameleon Pictures 24
Famous German Spy Formula 24
Vanishing Inks 24
Emergency Type Inks 25
Spirit Writing 25
A Magic Picture 26
A Magic Picture in Colors 26
Magic Blue Writing Paper 27
Magic Black Writing Paper 27
Writing that Disappears by Magic 28
Salt Writing 28
"Old Glory" Appears by Magic 28
The Mysterious Picture 28
The Answering Picture 28

CHAPTER IV LUMINOUS PAINTS
Phosphorescent Glowing Paint 30
Orange Luminous Paint 30
Green Luminous Paint 30
Green Luminous Paint 31
Blue Luminous Paint 31
Blue Luminous Paint 31
Violet Luminous Paint 31
Deep Yellow Luminous Paint 31
A Few Comments 32
Red Luminous Paint 32
An Easy Luminous Paint 33
A Ghost Show Routine 33

Ghost Card 37
Ghost Coin 38
Spirit Answer 38
Mysterious Dice 38
Mystery Magazine 39
Mentally Speaking 40
Evil-Eye 41
Spooky Lights 41
Apparitions 42
Ghostly Faces 43
Ectoplasm 43
Seeing Ghosts 44
The Escaping Shadow 45

CHAPTER V. LUMINOUS INKS
Luminous Ink 47
Fluorescent Inks 48
Luminous Paste Inks 48

CHAPTER VI. DAUB
Magic Methods with "Daub" 50
Magic with Daub 52
Location 53
Take A Card 53

CHAPTER VII. MAGIC WITH FLASH PAPER
Flash Paper 54
Colored Flash Paper 55
Colored Flash Paper 55
Dove Pan 56

CHAPTER VIII. FIRE MAGIC
Blue Fire 58
Red Fire 59
Green Fire 60
Yellow Fire 60
White Fire 60
Colored Flames 61
Water Starts A Fire 61
Ice Starts A Fire 62
The Exploding Hammer 62
Magic Sparks 62
Flash Powders 63
Serpent's Eggs 64

Some More Snakes 65
Burning Ice 65
Smoke Screen 65
The Hot Gin 66
Fired Liquid 66
Satan's Telegram 66
Think Fire 67
An Innovation 68
The Fire Bowl 68
Self-Lighting Candles 68
A Magic Picture 69
Lighting A Cigarette with Ice 69
Candles Which Go Out 70

CHAPTER IX. FIRE EATING
Blowing A Paper On Fire 72
Flames from the Mouth 72
"Eating" Fire 72

CHAPTER X. MAGIC WITH SLATES
The Scientific Secret 75
"The Spirits Know All" 76
A Daring Volta Method 78
Dead or Alive Discovery 79
A Number Test 80
The "Spirits" Correct A Mistake 82
A Few Remarks 83
A Mind Reading Number Divination 84
Another "No Flap" Method 87
Spirits from China 88
Suffice To Say 89
Colored Chalk 89
For Children Only 89
Additional Ideas 90
One of the Three Little Pigs 90
The Rising Card 91
Novelty Balloon Stunt 91
Magic Hair Restorer 91
Other Ideas 91
Some Other Slate Writing Techniques 92

CHAPTER XI. CHEMICAL COCKTAILS
The Magic Soda Fountain 94
Sensational New "Beer Powder" 96
The Magic Soda Fountain 96

CONTENTS

CHAPTER XII. CHEMICAL STUNTS

A Chemical Smoke Screen 102
Magic Ice 102
How to Make Bones Elastic 103
How to Stretch an Egg 103
The Magic Handkerchief 103
Green Violets 104
An Obedient Egg 104
Water That Will Not Spill 104
The Mysterious Mothball 105

CHAPTER XIII. PROFESSIONAL MAGIC

Patriotic Liquids and Silks 106
Magic Dyeing 107
Califaction of Handkerchiefs 109
Mystic Papers 109
The Bewitched Paper 110
Stage Size Out to Lunch Trick 111
The Chemical Garden 112
Magic Garden Seeds 114
Dry or Wet 114
Anti-Climax 114
Flaming Beauty 114
Reading Sealed Envelopes 115
Another Liquid—??—Solid 115
Writing with Fire 115
Spirit Photograph 116
The Spirits Reveal 116
Flash Pot 117
Dark Eyes 117
Revelation 119
Prediction 120
Misdirection 121
Bottle of Smoke 122
Production of Three Large Balloons from a Paper Cone ... 122
Another Balloon Production from Pape Cone 123
And More Balloons 123
Think! Ink! 124
Clock Reaction—Yellow 126
Clock Reaction—White 126
The Blackslider 126
The Money Maker 127
Magic Flowers 128

The Smoking Pipes 129
Smoke in the Bell Jar 130
The Atomic Bomb 130
The Magic Pipe 131
The Pirate Knife 131
The Spirit Glass 132
Magical Rainbow 133
The Smoking Hands 133
Gravatt's Mental Mystery 134
Ink and Water Columns 135
Sands of Enchantment 137
The Trick without a Name 138
How to Treat Rope Ends 140
Frozen Smoke 141
Magnet-Tizo 142
A Flash Paper and Match Stunt 144
Proding Two Flashes of Flame 144
A Self-Lighting Candle 145
Secret Message 146
Balloon "Skullduggery" 146
Floating Metal Disc 147
Pearls of Buddha 148
Magic Metal 148
Magic Dye 149
Ideas for Routines with Chemical Magic 149
Chemical Substitution 150
Ink—??? 150
Color Changes 151

CHAPTER XIV. CHEMICAL TIPS FOR BETTER TRICKS

Daub 152
Metallic Daub 153
Daub Substitutes 153
Milk Pitcher Magic 153
That Egg "Again" 153
Beverage Substitutes 154
Flesh Paint 154
Wax Substitutes 154
For Slicking Cards 155
Roughing Fluid 155
Pip Paint 155
Silk Cleansing Compounds 156
Cleaning Feather Flower 156
Removal of Stains from Cards 156

CONTENTS

Fire-Proofing 157
Stains from Metal 157
Cement for Glass 157
Diachylon 157

CHAPTER XV. CHEMICAL JOKES AND NOVELTIES

Exploding Matches 159
Exploding Matches 160
Exploding Matches 160
Diabolical Candles 160
Trick Matches 161
Sugar—??? 161
Serpent Matches 161
Vanishing Ink 162
Sparkling Matches 162
How to Make A Fuse 162
Magic Cigarettes 162
Diabolical Bombs 163
Chameleon Powders 164
Living—??? 164

Chapter 1

Chemical Magic With Liquids

Water to Wine

Prepare two glasses by putting in one a few drops of a solution of phenolphthalein. In the second glass half-filled with water dissolvc a small amount of sodium carbonate.

Pouring the water into the first glass will turn the clear liquid red, appearing as wine.

A Magic Pitcher of Grenadine

Fill a pitcher which is not transparent with water. Arrange five glasses around the pitcher prepared by placing

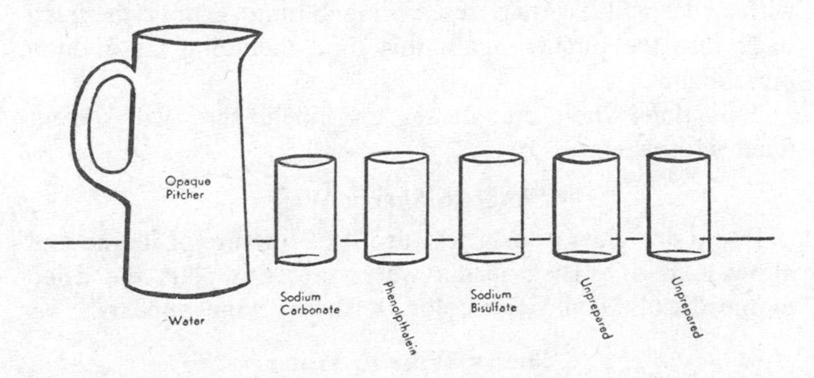

in one a small amount of sodium carbonate, in another a few drops of a phenolphthalein solution, in the third a small amount of sodium bisulfate. The remaining two glasses are left unprepared. This serves to complicate the effect to further confuse the spectator. Dissolve the dry chemicals with a few drops of water. Remember which glass contains the sodium bisulfate.

Fill all the glasses with water from the pitcher. Then pour the contents back into the pitcher, except the one containing the sodium bisulfate. This time when you fill the glasses they will appear filled with grenadine.

Now empty all of the glasses into the pitcher again, this time including the sodium bisulfate solution. When you fill the glasses for a third time, it will be clear water.

A Magic Pitcher of Bluing

Arrange a pitcher and glasses again as in the preceding effect. In one glass place sodium ferrocyanide, in another ferric ammonium sulfate and in another sodium carbonate.

Fill all the glasses with water from the pitcher and empty them back except the one containing the sodium bicarbonate. Filling the glasses for a second time causes them to be filled with a deep blue liquid resembling bluing. Empty them all back into the pitcher again this time including the sodium bicarbonate.

This time when the glasses are filled, they will appear filled with water.

Water to Sherry Wine

Into a dry glass pour a few drops of tincture of iodine and allow it to dry. By pouring water into this glass the dried iodine dissolves and the color of sherry wine appears.

Sherry Wine to Water

Prepare a glass half-filled with water and dissolve a small amount of sodium thiosulfate in it. This colorless solution can be called water and if poured into the sherry wine of the preceding effect, the wine will instantly turn to clear water again.

A Magic Pitcher of Wine and Bluing

Fill a pitcher which is not transparent with water. Arrange four glasses prepared by placing in the first sodium carbonate, in the second a few drops of a phenolphthalein solution, in the third some ferric ammonium sulfate and in the fourth sodium ferrocyanide.

Fill the glasses about half-full of water from the pitcher. (The pitcher should contain just enough water to fill each of the four glasses half-full.) Pour the contents of the first two glasses into the pitcher. Fill them up again and two glasses of wine appear.

Repeat this with the contents of the third and fourth glasses and two glasses of bluing appear.

Pouring Red, White and Blue from the Same Pitcher

Fill three glasses two-thirds full of water. In the first dissolve sodium salicylate. In the second dissolve strontium chloride. In the third dissolve sodium ferrocyanide. In a clear glass pitcher dissolve a small amount of ferric ammonium sulfate. Pour some of this solution in the pitcher into each of the three glasses.

The liquid in the first glass will turn red, in the second it will turn white and blue in the third.

Pouring Red, Yellow and Black from the Same Pitcher

Fill three glasses each one-half full of water and arrange them in a row. In the first glass dissolve a small amount of tannic acid, in the second some sodium bisulfite and in the third some sodium salicylate. Stir each glass until the solid is dissolved.

In a clear glass pitcher full of water dissolve some ferric ammonium sulfate. Stir until completely dissolved. Pour from the pitcher into each of the three glasses. In the first a black liquid will appear, in the second a yellow and in the third a red. This makes a very pretty series of colors.

Pouring Blue, Gold and Red from the Same Pitcher

Fill three glasses each one-half full of water and dissolve in the first a very small amount of sodium ferrocyanide, in the second glass some sodium bisulfite and in the third a small amount of sodium salicylate.

In a clear pitcher dissolve some ferric ammonium sulfate. Now pour from the pitcher into each of the three glasses. In the first a blue color results, in the second a golden yellow and in the third a red.

Pouring Nine Different Colors from the Same Pitcher

Arrange nine glasses in a row. In the first place a small amount of ferric ammonium sulfate and an equal amount of tannic acid.

In the second put a small amount of strontium chloride and an equal amount of sodium carbonate.

In the third put a small amount of sodium salicylate and an equal amount of ferric ammonium sulfate.

In the fourth repeat with cobalt chloride and sodium carbonate.

In the fifth repeat with calcium oxide and a few drops of a phenolphthalein solution.

In the sixth repeat with ferric ammonium sulfate and sodium carbonate.

In the seventh put a small amount of sodium ferrocyanide and an equal amount of ferric ammonium sulfate.

In the eighth repeat with cobalt chloride and sodium ferrocyanide.

In the ninth place some sodium ferrocyanide and ferric ammonium sulfate.

Now fill each glass from a pitcher of water, stirring each one slightly, after you have filled them all. The water will turn a different color as it goes into each glass. The effect is startling as well as beautiful.

Water to Milk and Milk to Ink

Dissolve some ferrous ammonium sulfate in a glass three-fourths full of water. In another glass place an equal amount of strontium chloride and add a few drops of water to dissolve the solids. In the third glass dissolve in a few drops of water an equal amount of tannic acid. Next pour the solution in the first glass into the second and the result will be milk. Pour the milk into the third glass and ink will result.

Wine to Beer

To make the wine, mix enough potassium chromate with a glass of water to give it a yellow beer color. Next mix in all the sodium bicarbonate that will dissolve in the glass of water, and last mix in a few drops of an alcoholic solution of phenolphthalein so that the mixture turns a dark red wine color. You will also need a weak solution of hydrochloric acid.

Use your regulation milk pitcher and fill the outside compartment with the acid solution. Inside (within the celluloid chamber) put the "wine." Be careful that the two solutions don't get a chance to mix until you are ready to perform the mystery. You must have a container that will hold all the liquid that is in the pitcher. If you have a large pitcher fill it only partly full, because once the two liquids mix, the "wine" will change to "beer."

The acid turns the red to clear, revealing the yellow of the potassium chromate, and the acid also acts on the sodium bicarbonate making foam.

This makes a good conclusion follow-up for any wine and water effect.

A Chemical Clock

Prepare in several glasses a series of solutions of sodium thiosulfate of decreasing strength; for example, place in the first glass ten parts, by weight, of the sodium thiosulfate; in the second glass eight parts, by weight, etc. Fill each glass half full of water and stir until the solid is dissolved.

Prepare a solution of sodium bisulfate, six parts, by weight, in water and add an equal portion of this to each glass. A milky precipitate will appear after a few minutes, but it will form in the different glasses in different lengths of time according to the concentration of the thiosulfate solution. A little experimenting makes it possible to create the appearance of the white precipitate within a specific time.

A Synthetic Cow

Fill a glass one-fourth full of water and add eight parts, by weight, of sodium thiosulfate. Stir until completely dissolved.

In another glass one-fourth full of water dissolve an equal amount of sodium bisulfate.

Pour the bisulfate solution into the glass containing the thiosulfate. Let it stand for about five minutes. A white "milky" precipitate will form, giving the solution the appearance of milk.

This effect is especially good for exhibition because the solution may be mixed behind the scenes and a glass of apparently clear water can then be set before the audience and after a while it will gradually turn milky. The milk mysteriously appears without anyone touching or going near the glass.

Conjuring Red from Blue

Fill two glasses each half-full of water. In the first glass place two or three drops of a solution of phenolphthalein. In the second a small amount of ferric ammonium sulfate and an equal amount of sodium ferrocyanide. Stir both glasses slightly until dissolved.

Prepare a third glass by placing a spoonful of sodium silicate solution and two teaspoonfuls of water in it. Stir slightly to mix.

Now take the first glass containing the colorless liquid in one hand and the second glass containing the blue liquid in the other hand. Pour them both at once into the third glass containing the silicate solution. A glass full of bright red liquid will result.

Water to Wine to Ink to Water

Fill a pitcher (not transparent) with water. Prepare five glasses in the following order. In the first place ten drops of a strong solution of iron chloride, in the second glass place two drops of a strong solution of ammonium sulfocyanide, in the third place twelve drops of a strong solution of ammonium sulfocyanide, in the fourth place twelve drops of a tannic acid solution and in the fifth glass fill it one-half full of an oxalic acid solution.

Fill the first glass from the pitcher with water. Empty it back into the pitcher. Then fill the second glass with wine appearing. Empty back into the pitcher and fill the third glass. Wine appears again. Empty back into the pitcher. Fill the fourth glass and ink appears. Fill the fifth glass with clear water appearing and empty back into the pitcher. Finally fill all five glasses from the pitcher with clear water resulting.

Pouring Ink and Milk from the Same Pitcher

In one glass place some tannic acid and in a second twice as much strontium chloride. Add about a teaspoonful of water to each glass and stir until dissolved.

In a small pitcher dissolve some ferric ammonium sulfate in water.

When ready to exhibit, pour half this liquid (from pitcher) into each glass. The liquid in the first one will become intensely black like ink and that in the second will become milky white.

Another Water to Wine

Dissolve a small amount of ferric ammonium sulfate in a pitcher of water. Obtain two glasses. Leave the first unprepared and in the second place a few drops of a sodium salicylate solution. Water poured into the first glass will appear as water and when poured (from pitcher) into the second glass, the color of a deep wine will appear.

Character and Knowledge

A clever double color change in which a clear solution is shown, but to which a second clear solution is added, causing

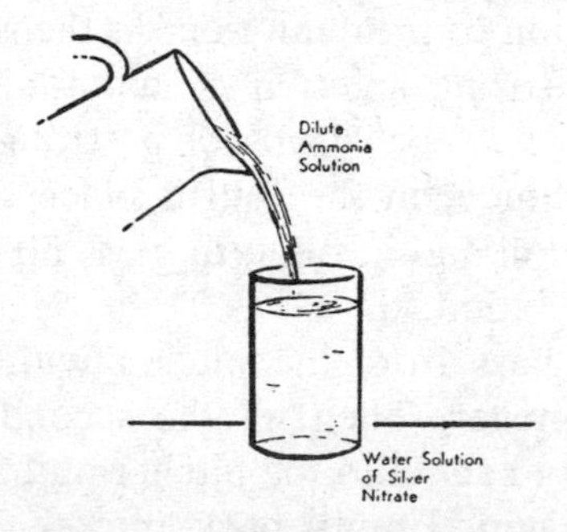

a brown cloudy color, but upon the addition of more of the second solution, this color fades, leaving a clear solution again.

Dissolve one-half gram silver nitrate in four-hundred-and-fifty cc. water. Add a few drops of an ammonia solution. A brown coloration will form. Adding more ammonia produces a darker color, until all of a sudden the solution becomes clear as crystal.

Several ways of presenting the phenomenon may be used. One is to declare that the original solution is the uninformed character of a young man. The acquisition of knowledge sometimes causes darkness and fear, but sufficient knowledge causes all to become clear.

Fugitive Color

This is a double color change which comes as something to surprise the audience.

In a glass place one cc. of a concentrated solution of sodium hydroxide. The quantity is too small to cause any suspicion.

In a pitcher place several hundred cc. water. Tap water will do. In this put one cc. of phenolphthalein solution and enough sulfuric acid so that one glassful of the solution will just neutralize the sodium hydroxide solution in the tumbler.

Fill the glass up with water from the pitcher slowly. As you do this, the liquid in the pitcher turns to red. When this happens, pause a moment and then continue pouring. As you fill the glass, the red liquid becomes colorless again.

As a suggestion for possible routines, after having seemingly produced a glass of wine from a pitcher of clear water, you pause and remark that perhaps your victim would prefer water. Hesitate a moment and proceed to fill the glass. As the glass is filled, it is noticed that suddenly the same wine has again become water.

Another suggestion: Calling upon one of your audience, get him to say the color of the liquid you are pouring. All will agree that the liquid in the pitcher is water. The victim insists that what you are pouring is red. You may prolong the argument insisting that it is colorless, until the change occurs, when you can prove you are right.

If any trouble is encountered in preparing the solutions for neutralization, a consultation with your local druggist will help.

Water to Milk

Fill two glasses half-full of water and dissolve in one some manganous sulfate. In the other dissolve an equal amount of sodium carbonate.

Now pour the contents of one glass into the other and a thick white precipitate will be formed, making it seem as though the glass were full of milk.

Water to Wine and Wine to Milk

Dissolve in a glass three-fourths filled with water some sodium carbonate. Stir to dissolve the solid. In another glass put a few drops of a solution of phenolphthalein, and in a third place a small amount of strontium chloride and an equal amount of sodium bisulfite. Pour the solution in the first glass into the second. The appearance of wine makes its presence. Pour the wine into the third glass to cause the change of the wine into milk. Stir vigorously for a few seconds with your wand to cause this change.

A Magic Pitcher of Milk and Ink

Obtain a pitcher and four glasses. In the first glass place some manganese sulfate. In the second an equal amount of strontium chloride and in the third a lesser amount of tannic acid. Prepare the fourth glass by placing a small amount of ferric ammonium sulfate. Add a few drops of water to each glass (enough to dissolve). Fill the pitcher with enough water to fill each glass.

Pour from the pitcher into each glass, and then pour the contents of the first two glasses back into the pitcher. Fill them again from the pitcher and this time they will appear filled with milk. Now pour the remaining two glasses back into the pitcher and when you refill them, the liquid will be ink.

Be sure to use a non-transparent pitcher.

Chemical Color Chase

Arrange five dry glasses in a row. Put in the first glass a small amount of ferric ammonium sulfate, in the second glass an equal amount of sodium salicylate, in the third glass an equal amount of sodium ferrocyanide, in the fourth glass a spoonful of sodium silicate solution, and in the fifth a few drops of a solution of phenolphthalein.

Fill the first glass with water and stir to dissolve the solid. Pour contents of the first glass into the second, the second into the third, and so on, and a change in color will occur. Stir each time.

Water to Wine

Prepare four empty glasses in the following manner. In the first put five drops of phenolphthalein solution, the second leave empty, in the third put five drops of a phenolphthalein solution and in the fourth put ten drops of a strong tartaric acid solution. Dissolve a small amount of sodium carbonate in a pitcher of water which has enough to fill the four glasses.

Fill the glasses with water from the pitcher. The first and third glasses will appear as being filled with wine, while the second and fourth glasses will appear to have nothing but water in them. Empty all the four glasses back into the pitcher and when the glasses are filled up for the second time, they will all be filled with water.

Multicolor with Flash Finish

A series of color productions in the modern manner with a colorful flash finish.

Obtain nine glasses empty and dry. In the first place five drops of a twenty percent aqueous sulfuric acid solution, in the second five drops of a phenolphthalein solution, in the third ten drops of a thymolphthalein solution, in the fourth five drops of an orthocresolphthalein solution, in the fifth five drops of an orthocresolphthalein solution and ten drops of a thymolphthalein solution, in the sixth five drops of a paranitrophenol solution, in the seventh five drops of a twenty percent aqueous sulfuric acid solution, in the eighth ten drops of a lead acetate solution and in the ninth some "beer powder."

Pour from the pitcher filling the glasses in the following order: starting with the first and through to the seventh. Fill glass number eight by pouring contents of the seventh glass into the eighth. This produces "milk." Return glasses six, five, four, three and two to the pitcher. Finally return the first glass back to the pitcher to restore the entire solution to water.

A flashy finish may be had by filling up glass number nine with the water and ending up with a glass of foaming "beer." CAUTION—DO NOT DRINK ANY OF THE SOLUTIONS OR THE "BEER."

Water to Beer

In a tall Pilsener glass place in the bottom the following: ten cc. of an alcoholic solution of soap bark, a small portion of sodium carbonate, and enough carmel to color the solution to resemble beer. In a pitcher of water place ten per cent sulfuric acid. Pouring the water into the glass will result in a foaming glass of beer. CAUTION—DO NOT DRINK.

Rainbow Colors

Perhaps by this time the reader will have realized that the various color changes explained in the previous effects are the end results of specific chemical reactions between chemicals that react only to produce those colors under certain isolated conditions.

This next monograph gives several examples with which the "experimenting magician" with a flair for research will be able to devise several of his own effects.

The authors of this book on CHEMICAL MAGIC have listed

as follows a few further varieties of each color of the rainbow.

Red: (1) To a solution of potassium chromate add a few drops of an aqueous solution of sulfuric acid. (2) To a solution of sodium sulfocyanide add a solution of ferric ammonium sulfate. (3) A solution that appears yellow at first and slowly changes to red may be caused by adding to a solution of mercuric chloride a solution of potassium iodide.

Yellow: (1) Add an aqueous solution of lead acetate to an aqueous solution of potassium chromate. (2) By mixing a solution of sodium bisulfite with a solution of ferric ammonium sulfate a yellow color will appear.

Blue: (1) Add a dilute aqueous solution of potassium ferrocyanide to a dilute aqueous solution of ferric chloride. (2) To a solution of ammonium hydroxide add slowly a solution of copper sulfate.

Orange: (1) A solution prepared by first mixing together solutions of sodium hydroxide and sodium dichromate should be added to a solution of lead acetate.

Violet: (1) Allow crystals of iodine to dissolve in carbon tetrachloride. (2) Allow some potassium permanganate to float on top of some water in a tall cylindrical container.

Green: (1) Add some sodium hydroxide solution to a solution of nickel chloride. (2) The following solutions mixed together will produce a green color—ferric chloride, lead acetate, sodium dichromate and potassium ferrocyanide. (3) to a very small amount of arsenious acid solution add a few drops of ammonium hydroxide. To this solution add an aqueous solution of copper sulfate.

Crème De Menthé

Boil some red cabbage leaves for thirty minutes. Strain this solution and allow to cool. In an opaque pitcher place some of this solution.

Prepare a glass by putting a few drops of ammonium hydroxide.

When you pour the solution from the pitcher into the glass, a solution resembling "Crème De Menthé" results. The effect can be achieved more realistically by using the correct glass in which this ever-popular drink is served.

By Command—Water to Ink

Prepare a solution of potassium iodate by dissolving one gram of the chemical in 500 cc. of water.

The second solution required for this "magical" effect is prepared by dissolving 0.2 grams of sodium bisulfite in 450 cc. of water and adding 3 cc. of a solution of 1 cc. of sulfuric acid in 20 cc. of water.

Mix equal quantities of the two solutions. In about 20 to 30 seconds the liquid will suddenly turn to the color of blue-black ink.

By careful timing, the magician can cause a seemingly miraculous effect by causing "water" to change to "ink" by command.

A Liquid from Solids

Mix together equal portions of lead acetate and sodium sulfate. Shake this mixture and it will turn to a liquid.

The chemicals used in this magical effect must be very finely powdered. You must be certain to powder them separately, not together.

Chemical "Ice"

Use undiluted sodium silicate (water glass) in this next magical demonstration. To a portion placed in a suitable container add a small amount of hydrochloric acid and shake. The two liquids will form a solid mass and have much the same appearance as ice.

Chameleon Mineral

Grind together equal portions of manganese dioxide and sodium hydroxide. Heat this mixture in a porcelain crucible strongly for 30 minutes. Allow to cool and to some of the black residue that remains add some water. A green solution results.

Add a few drops of sulfuric acid to the green solution and it will change at once from green to red.

That Chameleon Mineral Again

Place some of the green solution from the previous demonstration into a glass and leave it stand by itself. As time passes it will slowly change to a bottle-green, then violet and finally will become a crimson-red.

Disappearing Whisky

Prepare an empty whisky bottle by dissolving enough iodine in water to resemble the color of what appears to be whisky.

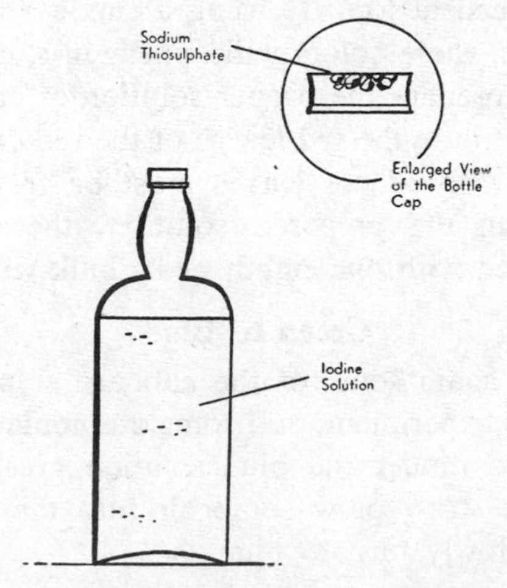

Prepare the cork by boring out a small hole in which is placed a few crystals of sodium thiosulfate. When the bottle is inverted momentarily, the crystal of sodium thiosulfate mixes with the iodine solution (whisky??) and turns colorless immediately, resembling water.

Solid to Liquid

Mix together equal portions of cold, concentrated solutions of calcium chloride and sodium carbonate. Stir gently but thoroughly and the solution will resolve into a jelly-like solid.

Blue to Red, Green, Crimson or Purple

In each of two glasses place a solution made by boiling in water red cabbage leaves for thirty minutes. This solution will be a dark blue color upon concentration. To one glass add a drop of sulfuric acid. The solution will change to crimson color. To the other add a drop or two of an ammonia solution and the visible result will be a bright green.

If you carefully allow a single drop of sulfuric acid to fall down the sides of a glass containing the cabbage leaf solution, crimson will appear at the bottom of the glass, purple in the middle and green at top. By using a single drop of the ammonia solution, these colors will appear in a reversed order.

In order to prepare the proper solution of cabbage leaves, it is important to use the red leaves of the red cabbage known as *Brassica Rubra.* These leaves must be freshly removed. After decanting the prepared solution, the clear solution should be mixed with one-eighth of its bulk with alcohol.

Green to Blue

Using once again some of the cabbage solution prepared in the previous experiment, add some ammonia solution. Add only enough to render the blue solution green. Through a paper or glass straw blow some air into the solution. The solution will slowly turn to blue.

Blue to Red

And if you continue to exhale air into the blue solution from the previous demonstration, it will turn red.

Chapter **II**

Fun With Dry Ice

Dry ice is solidified carbon dioxide gas and can bc used to perform many amusing experiinents. One word of caution . . . . . DO NOT HANDLE THE PIECES OF DRY ICE IN THE HAND AS YOU WOULD OTHER OBJECTS, FOR SERIOUS AND DANGEROUS BURNS ARE LIABLE TO RESULT. It should be handled with extreme care, preferably forceps.

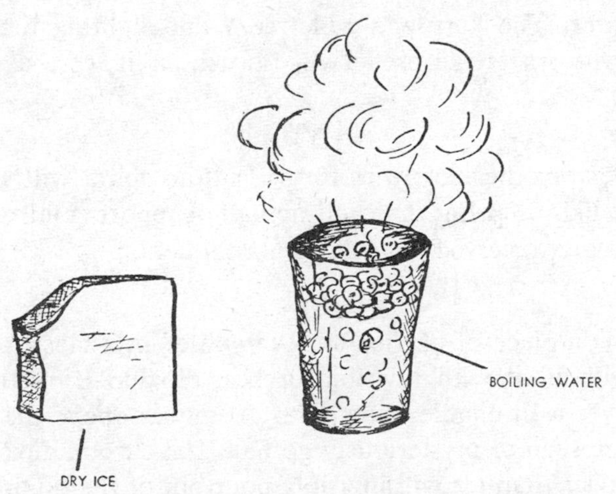

I

When dry ice is placed in water, the water will appear to be boiling, which will continue as long as the dry ice remains.

II

A small particle of dry ice placed in a balloon will give off enough gas to inflate it.

III

Striking a piece of dry ice with a hammer creates such great heat and pressure that the dry ice disappears immediately.

IV

By packing in dry ice for several hours: meat, flowers, rubber bands, fruits, soft candy, etc. will become so frozen that they will fly into countless pieces when struck on a hard surface.

V

Prepare a cylindrical tube of paper and pack dry ice around it. Pour kerosene into it with a piece of string down the center. The kerosene will freeze and lighting the string (wick) the frozen kerosene will burn as if it were a regular candle.

VI

Solder or other soft wire formed into a spiral will actually behave like a spring that will actually support small weights after having received the "dry ice treatment."

VII

Allow a piece of dry ice to "evaporate" in a glass tumbler. This will fill it with invisible carbon dioxide. Inverting the glass over a lit candle about 8 to 10 inches above the flame. it will be seen to mysteriously go out. The carbon dioxide be ing heavier than air will invisibly pour out of the tumbler and mysteriously smother out the flame.

VIII

Mercury can be frozen hard enough by dry ice that it can be used to hammer nails into wood.

Chapter III

Invisible Inks

The subject of invisible inks (sympathetic inks) and of magic writing is one that covers much material. By way of explanation, a sympathetic ink is one that becomes visible upon application of certain media both physical and/or chemical. They can also be made to remain permanently visible or to again vanish into the realm of invisibility.

The inks look like clear water and when dry are invisible on most soft white papers. Write with a clean pen. In compounding these inks always use distilled water.

The formulae and effects explained here in these pages are but a few of the many available. Only the imagination of the person using these secrets will devise the many mysterous deviations possible.

Red

I

Write with fifteen grains of potassium iodide dissolved in one ounce of distilled water. Sponge over with a solution of twenty grains of mercury bichloride dissolved in one ounce of water.

II

Write with a weak solution of copper nitrate. When the writing is exposed to mild heat it will become visible.

III

Writing with a strong alcoholic solution of phenolphthalein is invisible when dry. Exposing it to the fumes of a strong solution of ammonia causes the writing to become invisible. As the ammonia fumes evaporate after removing the paper, the writing will gradually disappear. By breathing on the paper the writing will vanish almost immediately.

IV

Write with a weak solution of silver nitrate. Upon mild application of heat a rose-red color will appear.

V

Write with a ten percent solution of potassium ferrocyanide. Apply a fifty percent solution of iron tincture to produce a red color.

VI

Write with an aqueous solution of iron chloride. Upon addition of a solution of sodium sulfocyanide a visible red color will appear.

Blue

I

Write with a weak solution of cobalt chloride. When mild heat is applied it will become visible. Upon cooling it vanishes.

II

Writing with a cobalt nitrate solution and then wetting with a weak solution of oxalic acid, a blue color will appear.

III

Write with an aqueous solution of copper sulfate. Make visible by sponging with a solution of iron chloride.

IV

Write with a solution made by dissolving fifteen grains of copper sulfate dissolved in one ounce of water. Sponge with a solution of fifteen grains of ammonium hydrate in one ounce of water.

Black

I

Write with a solution of one part sulfuric acid mixed with ten parts water. Writing appears when the paper is carefully heated.

II

Write with a fifteen percent solution of ammonium hyper-chlorate. Heat the paper carefully to develop the writing.

III

Write with a solution made by dissolving twenty grains of iron sulfate dissolved in one ounce of water. Develop the writing by sponging with a solution of five grains of tannic acid dissolved in one ounce of water.

IV

Writing with a solution of starch boiled in water will turn black when treated with tincture of iodine.

V

Writing with a strong solution of mercurous nitrate and with exposure to ammonia fumes causes black writing to appear.

Brown

I

Write with fifteen grains of copper sulfate dissolved in one ounce of water. Sponge with ten grains potassium ferro-cyanide dissolved in one ounce of water.

II

Write with a 1:1000 solution silver nitrate. When exposed to light the writing becomes visible.

III

Prepare and write with a solution of lemon juice to which

has been added a slight amount of citric acid. Warm the paper slightly to cause the writing to develop. Do not overheat as the writing will not disappear upon cooling.

IV

Vinegar when allowed to dry and then exposed to mild heat will result in brown writing becoming visible.

Yellow

I

Writing with a solution made by dissolving equal parts of copper sulfate and ammonium chloride will result in yellow writing to appear upon application of mild heat.

II

Write with twenty grains mercury bichloride dissolved in one ounce of water. Sponge with ten grains sodium hydroxide dissolved in one ounce of water.

III

Write with a very dilute solution of copper perchloride and apply mild heat to make the writing visible.

IV

Write with an aqueous solution of potassium bromide. Apply a similar solution of copper sulfate to produce yellow writing.

Green

I

Write with an aqueous solution of sodium chlorate. Develop the writing with an aqueous solution of copper sulfate.

II

Apply an aqueous solution of potassium ferrocyanide to writing prepared with a similar solution of cobalt nitrate.

III

Write with an aqueous solution of copper chloride. Apply a similar solution of cobalt chloride to develop the writing.

IV

Write with a weak solution of nickel nitrate. Apply mild heat to make the writing visible.

V

Write with a weak solution of nickel chloride. Apply mild heat to develop the writing.

VI

Write with the following solution: Digest one ounce of cobalt oxide in four ounces of nitro-muriatic acid by the aid of very mild heat. Then add one ounce of sodium chloride dissolved in sixteen ounces of water. Writing will appear if mild heat is applied.

Violet

I

Writing with an aqueous solution of carbolic acid (phenol) and then applying an aqueous solution of iron chloride will cause visible writing.

Water Creates Visibility

Writing with a solution of bismuth nitrate, will, when dry, be invisible. But on wetting the paper with water, the writing will appear in dense white letters.

Light Creates Visibility

Write with a weak solution of silver nitrate. Allow the characters to dry and they will be completely invisible. Do not expose the paper to light until ready to cause the writing to appear.

Chameleon Pictures

First sketch a landscape in waterproof ink. It should resemble either a winter scene or a mountain district. When complete and throughly dry apply invisible inks as follows:

SKY and FROZEN LAKES ....cobalt acetate
COTTAGE THATCHES ....copper chloride
FLOWERS ......copper chloride, copper perchloride, etc.
TREES ....cobalt chloride

All of these solutions should be applied with various degrees of concentration to effect variations in color intensities. When dry, the picture will appear bleak and drab as the applied chemicals are invisible when dry.

When ready to cause the picture to assume its beautiful proportions as any such intended scene should, subject it to mild heat and the various colors will appear in all their glory.

Famous German Spy Formula

Write with one ounce of linseed oil and twenty ounces of ammonia water (not household kind) in one-hundred ounces of distilled water. Shake well when using. This ink will become visible when the paper is dipped in clear water and will disappear when the paper becomes dry. On re-wetting, the writing will return.

Vanishing Inks

I

Write with fifteen grains of potassium iodide and fifteen grains iodine crystals dissolved in one ounce of water. Use the same as you would ordinary ink. In several days the writing should disappear.

II

A quickly vanishing ink (red) can be made by dissolving a small quantity of phenolphthalein in strong ammonia water. Writing with this ink disappears a short time after it has been exposed to the air a few moments.

Emergency Type Inks

When the usual chemicals found in the chemist's laboratory are not handy, many substances found in the kitchen can be used as a very satisfactory invisible ink. They require but moderate heat to develop the writing.

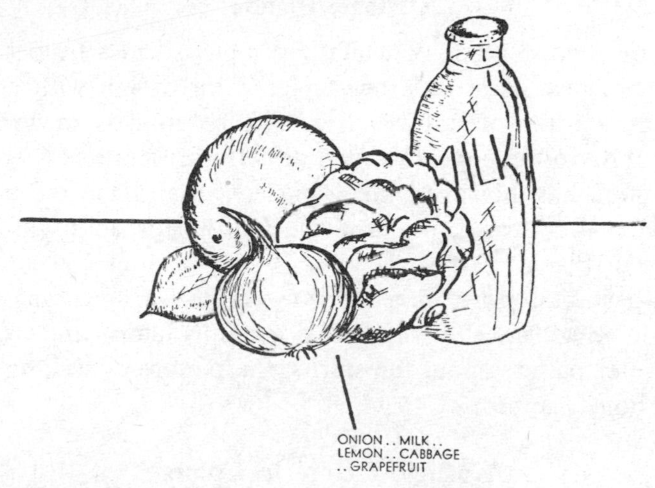

ONION . . MILK . .
LEMON . . CABBAGE
. . GRAPEFRUIT

Several of such substances are as follows:

Lemon juice	Onion juice
Leek juice	Cabbage juice
Milk	Artichoke juice
Buttermilk	Grapefruit juice

Spirit Writing

Using sheets of good quality white paper, spray or brush them with the following solutions:

Sodium ferrocyanide solution
Sodium salicylate solution
Tannic acid solution

Or paint alternate streaks of the above chemicals on one sheet.

Dry the prepared sheets of paper and press them flat.

Using a clean pen, write upon the prepared sheets with a solution of ferric ammonium sulfate. The writing appears a different color on each sheet or changes color as you write across a sheet prepared with the three separate chemicals.

A Magic Picture

Using a good grade of white paper tacked to a frame, and clean brushes, paint or draw a picture or design with a five percent solution of sulfuric acid. This solution is so weak it will not harm the paper, but will dry into invisibility. A chemical action has, however, taken place so that when the paper is heated or warmed, the area touched by the acid will turn brown or black.

Suggestion for use might revolve about a spirit message (placed in an envelope against a reading lamp) and as the performer patters about the spirits, the picture or writing will mysteriously appear.

A Magic Picture in Colors

Fasten a large sheet of a good grade of white paper to a board supported upon an easel.

Paint this paper with the following solutions, one at a time to avoid "running." Use a clean brush for each solution as follows:

Sodium ferrocyanide (blue)
Sodium salicylate (red)
Tannic acid (black)
or use
Potassium ferrocyanide (blue)
Potassium thiocyanate (red)
Tannic acid (black)

When dry, produce the colors "magically" by brushing or spraying with an atomizer a solution of ferric chloride.

Magic Blue Writing Paper

Mix equal amounts of sodium ferrocyanide and ferric ammonium sulfate. Rub these chemicals on a smooth white piece of paper. Shake off the extra powder and write using only water.

Some people will not believe that you can write with plain water, but just prepare some sheets of paper this way and let them try it. To further mystify them have a few untreated sheets to try beforehand.

Magic Black Writing Paper

Lay a sheet of ordinary writing paper on a smooth flat surface. Place equal amounts of tannic acid and ferric ammonium sulfate on it and proceed as outlined in the previous effect. Instead of blue writing as before, there will appear black words and lines.

Writing that Disappears by Magic

Prepare a solution of ferric ammonium sulfate and sodium ferrocyanide in water. Write on white paper with this freshly prepared ink.

Prepare a second solution of sodium carbonate in water. Wet a blotter with this second solution (do not allow audience to notice the dampness of the blotter.) Then blot the writing with this prepared blotter and the writing will vanish.

Salt Writing

Dissolve common table salt (sodium chloride) in water. Write with this solution on white paper using a clean pen. When the writing is dry, scratch over it with a soft pencil and the words you have written will show up plainly in dark lines.

"Old Glory" Appears by Magic

Draw a flag on a sheet of paper in light pencil lines, laying out the blue field and stripes. Now prepare the following solutions:

Sodium ferrocyanide dissolved in water
Sodium salicylate dissolved in water
Ferric ammonium sulfate dissolved in water

Paint the blue portion of the flag with the sodium ferrocyanide solution. The stripes which are to be red are painted with the solution, sodium salicylate.

Allow the drawing to dry thoroughly and then paint over the entire picture with the ferric ammonium sulfate solution using a small brush or piece of cotton. The flag will come out in the beautiful red, white and blue pattern of "Old Glory."

The Mysterious Picture

Using a small, soft brush, paint a picture on a sheet of paper with cobalt chloride solution. Dry the paper with very mild heat being very careful not to scorch it. When the paper is warm and dry, the picture will appear in blue.

To make it disappear again, breathe on it for a minute or two or hold it in the steam of a boiling tea-kettle. You can make the picture appear and disappear as often as you like.

The Answering Picture

This principle about to be described may be adapted in many ways in the revelation of an answer to a question routine presented by the performer. The modus operandi is illustrated in this simple comical effect, but remains in no manner changed when used as your imagination allows.

On one side of a sheet of paper (white), draw a picture of a boy riding a tiger. Use an ink prepared by dissolving a few crystals of copper sulfate in seven ounces of water. Expose

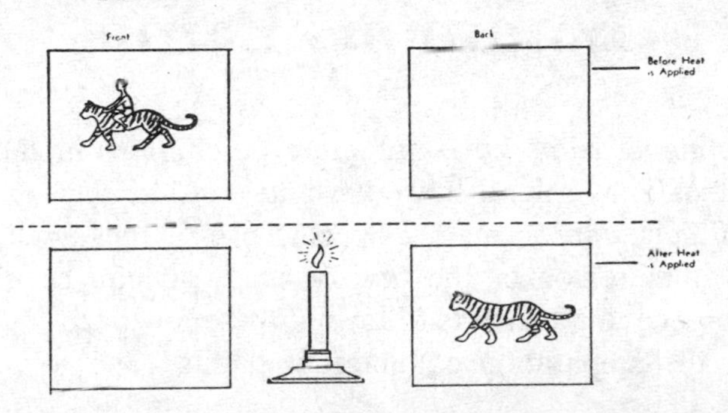

this to ammonia fumes until it turns blue. On the other side, draw a picture of a rather well-fed tiger (with no boy). Use an ink prepared by adding one ounce of concentrated sulfuric acid to five ounces of distilled water. When dry this second picture will be completely invisible.

Exposure to a source of heat will mysteriously cause the tiger and boy to disappear and the now not-so-hungry tiger to appear.

Chapter **IV**

Luminous Paints

Luminous paints have unlimited possibilities in the realm of magic. They might, however, find considerable use in the field of spiritualistic magic. Specific use of these will be left to the imagination of the performer. Ideas can be obtained by reference to the effects following these few pages of formulas for the preparation of luminous paints.

Phosphorescent Glowing Paint

Heat strontium thiosulfate for fifteen minutes over a strong flame, then for five minutes with a blow torch. It can then be used as a phosphorescent paint. To employ it as a paint, it is necessary to mix it with pure melted paraffin. Expose it to strong sunlight to initiate its phosphorescent properties. Painting with this paint will result in a mysterious glow of objects so treated.

Orange Luminous Paint

Varnish	23 parts by weight
Prepared barium sulfate	8.75 parts by weight
Prepared India yellow	0.5 parts by weight
Prepared madder lake	0.75 parts by weight
LUMINOUS CALCIUM SULFIDE	19 parts by weight

Green Luminous Paint

Varnish	12 parts by weight
Prepared barium sulfate	2.5 parts by weight
Green chromic oxide	2 parts by weight
LUMINOUS CALCIUM SULFIDE	8.5 parts by weight

Green Luminous Paint

Strontium thiosulfate	60 parts by weight
0.5% Acidified solution of bismuth nitrate	12 parts by weight
0.5% Alcoholic solution of uranium nitrate	6 parts by weight

The materials are mixed, dried, brought gradually to a temperature of 2,372 degrees F., and heated for about one hour.

Blue Luminous Paint

Varnish	21 parts by weight
Prepared barium sulfate	5.1 parts by weight
Ultramarine blue	3.2 parts by weight
Cobalt blue	2.7 parts by weight
LUMINOUS CALCIUM SULFIDE	23 parts by weight

Blue Luminous Paint

Calcium oxide (burnt lime) free from iron	20 parts by weight
Sulfur	6 parts by weight
Starch	2 parts by weight
0.5% Solution of bismuth nitrate	1 part by weight
Potassium Chloride	0.15 parts by weight
Sodium Chloride	0.15 parts by weight

The materials are mixed, dried, brought gradually to a temperature of 2,372 degrees F., and heated for about one hour.

Violet Luminous Paint

Strontium carbonate	100 parts by weight
Sulfur	100 parts by weight
Sodium chloride	0.5 parts by weight
Potassium chloride	0.5 parts by weight
Manganese chloride	0.4 parts by weight

The materials are mixed, dried, brought gradually to a temperature of 2,372 degrees F., and heated for about one hour.

Deep Yellow Luminous Paint

Strontium carbonate	100 parts by weight
Sulfur	30 parts by weight
Sodium carbonate	2 parts by weight
Sodium chloride	0.5 parts by weight
Manganese sulfate	0.2 parts by weight

The materials are mixed, dried, brought gradually to a temperature of 2,372 degrees F., and heated for about one hour.

A Few Comments

It is highly important in the preparation of the above formulas, that the ingredients used be as finely powdered as is possible. Do not forsake the use of pure chemicals for cost, as the final results will be lacking in satisfaction of luminosity.

To make a good flexible luminous paint, mix any of the above with a high grade of Aerated Varnish. If too heavy, thin them with turpentine. Use a good brush and allow 24 hours for the paint to dry. Do not hurry-up the drying process by use of artificial or mechanical sources of heat.

A coat of white shellac will protect the luminous surfaces. Since there are houses that have luminous paint already prepared, the necessary requirements to make your own can be avoided; it is recommended that the user of such paint first consider the purchasing of an already prepared luminous paint.

Haines' House of Cards, Cincinnati 12, Ohio manufactures a brand of luminous paint known as "Ghostlite" and can be obtained direct from this manufacturer or from any of the several established magic dealers in the country.

When used as the manufacturer recommends, "Ghostlite" will create "spooks" and will cause other objects to become visible in the dark with a misty-bluish, ghostly-like color.

Red Luminous Paint

Calcium sulfate	100 parts by weight
Willow charcoal	30 parts by weight
Sodium carbonate	2 parts by weight
Sodium chloride	0.5 parts by weight
Manganese chloride	0.2 parts by weight

Heat this mixture to red heat in a crucible over a bunsen burner. To make it into paint, mix it with spar varnish. Do not use water or glue as these will decompose the sulfides in the paint. If your luminous paint does not glow after exposure to a bright light it is probably because you did not secure the necessary high temperature in the preparation.

An Easy Luminous Paint

Barium sulfate .... 5 ounces
LUMINOUS CALCIUM SULFIDE .... 18 ounces

Grind the above in a mortar until a very finely divided powder results. After they are thoroughly mixed, stir them into 24 ounces of a clear four-hour varnish. Keep the can sealed when not in use. The surfaces to be painted must be thoroughly dry and clean.

A Ghost Show Routine

The lights slowly dim out. The theater is dark with only the exit lights on. A large luminous bell with swinging clapper appears at one side of the stage and moves to the opposite side where it vanishes. The bell rings on its trip across the stage. There is a sudden flash of light and four ghostly heads appear, one at a time, and float about the stage. They vanish one at a time. A luminous skeleton then appears and dances about the stage. It divides into three skeletons who continue with the dance. The three skeletons dissolve back into one skeleton which also fades away.

Two luminous ghostly forms fly out over the heads of the audience. The audience in turn experiences odd sensations. There is a feeling of rain . . . . . chains rattle . . . . . a flash of light and the ghostly forms disappear. The large bell ringing, slowly moves back across the stage and disappears. The house lights are slowly brought up.

Paraphernalia: The BELL is made of ⅛-inch plywood, 23 inches high. It is painted with two coats of luminous paint. Paint a flat black margin around the bell edges. The clapper on the back is separated a bit from the bell proper and pivoted by means of a nail. The clapper must swing back and forth easily. The back of the bell and clapper are painted flat black.

The GHOSTLY HEADS are made and handled like ordinary fans. Each has a handle fastened on the back. The fan proper is 24 inches high, and the handle is fifteen inches long, extending eight inches below the fan. The heads are of 1/8 inch plywood, painted with two coats of luminous paint, with flat black around the outside, on the handles and on the backs.

The THREE SKELETONS are the type usually found at Halloween. Take them apart, and paste them on heavy black cardboard. Give the white parts two coats of luminous paints. When dry, rejoin them with brass paper clips secured from any stationery store. The joints must move easily. Paint the backs of the skeletons black.

The GHOSTLY FORMS are made from two pieces of muslin about 18 inches wide and 6 feet long. These are fastened to two long fishing poles. The muslin should be tied to the poles with strong cord. About six inches of cord between the end of the pole and the muslin is sufficient. Paint the muslin with luminous paint.

The FLASH POTS are simple in construction. From any electrical supply store get a metal switch box with a porcelain socket mounted inside and connected to a long light cord. Remove the top from a fuse plug and screw into the socket. Place a *small* quantity of flash powder (formula will be found elsewhere in the book—See chapter VII, Fire Magic) into the fuse on top of the lead wire found inside the fuse. When the current is sent into the fuse, it is shorted and the powder explodes, causing a flash of fire and a great quantity of smoke.

The STAGE CURTAINS should be dark, as should the CYCLORAMA if one is used. This will increase the mysterious approaches of the act as black is so often associated with the supernatural.

The STAGE LIGHTS should be made from 4-inch squares of cardboard with large spots of luminous paint on them. These should be tacked to the scenery and in the footlights so the performer and his assistants on stage can see their exists and entrances in the dark.

The STAGE COSTUMES should be black robes and hoods that should cover the assistants completely.

The BELL can be any sound mechanism operated off-stage to imitate a ringing bell.

The GHOSTLY LIGHT ON PERFORMER can be created by a flashlight with green gelatin over the lens.

Performance: The performer enters. All lights, including stage lights are all the way up. The performer secretly carries the small flashlight with the green gelatin over the lens.

"Ladies and gentlemen, much has been said throughout the past seventy-five years regarding the return to earth of ghostly forms. It has been a popular sport to tell ghost stories, and many people agree that they have actually seen ghosts and apparitions playing eerie pranks. One man said he didn't believe in ghosts, but had been afraid of them all his life. If you feel that you have actually seen a ghost sometime in the past, I ask that you compare its appearance with any apparition that may materialize this evening, on this stage or out amongst you, the audience."

The lights slowly dim out. Music starts softly (J. Massenet's Elegie is an appropriate theme). The performer secretly steals the flashlight and throws the green beam of light on his face.

"Above all, do not let your imagination run away with you. In darkness, life was created, and it is in darkness that the unknown world reveals itself. In the light it is impossible to see the fragile forms that perform these mysteries. Darkness alone will cause them to be seen by the human eye, and even then only the most sensitive eyes will see all that takes place. In fairness to all gathered here this evening, I ask that you refrain from using flashlights. Be as quiet as possible, and no matter what happens, please remain in your seats."

The flashlight goes out. The next line is spoken in complete darkness. "I feel that we do have spirit friends with us. We shall soon see."

The large bell, with swinging clapper, appears at one side of the stage and slowly moves to the opposite side where it vanishes. The luminous side of the bell is toward the audience as it goes across the stage. An assistant back stage operates the bell-ringing mechanism.

The flash pot in the footlights is now set off. The four large ugly heads appear, one at a time, and float about the stage. Two assistants, with a head (fan) in each hand can accomplish this. The heads vanish one at a time by merely turning the luminous side of the fans away from the audience. In the beginning they are turned, one at a time so the luminous side comes toward the audience and are moved up and down as though floating in the air.

One assistant comes to the center stage holding the three cardboard skeletons with the black side towards the audience. The assistant turns one of the skeletons toward the audience and gives it to a second assistant who causes it to dance for a few moments. First assistant then turns the second skeleton and gives it to the third assistant who causes it to dance about. The first assistant then turns the last skeleton and makes it dance.

It appears to the audience that one skeleton has divided into three. The process is now reversed until all the skeletons have vanished. They seem to have dissolved into each other. As the skeletons appear or disappear they are brought to the center of the stage and held in front of the first assistant.

Second and third assistants now pick up the fishing poles with the luminous streamers and they are floated, quickly, over the heads of the audience. At the same time two stooges in the opposite upper balconies throw several small handfuls of rice into the air so it will settle down on the audience like rain. During this part there should be a rattling of chains back stage which stops when a second flash pot is set off in the footlights. With the flash the streamers vanish.

The large bell again slowly moves across the stage, with clapper swinging as the assistant back stage operates the bell-ringing mechanism again.

The curtain is lowered on a dark house and the house lights are slowly brought up.

Ghost Card

Spectator selects a card and the lights in the room are extinguished. Presently a large luminous card pip is seen floating around. Other pips appear and after floating about in the dark, the pips come together and form a ghostly image of the selected card. Here is a spectacular effect, simple and easy to make or present.

With luminous paint, make a large diamond on each of three pieces of cardboard four inches square. Black cardboard is preferred, although any color will answer the purpose. Expose the painted sides of the cardboards to a bright electric light for thirty seconds after which put the cardboards into your left coat pocket. Secretly place the three of diamonds on the bottom of a deck of cards and you are ready to present this effect.

Have spectator select a card. The selected card must be the three of diamonds, therefore it should be forced. Use your own method of forcing.

Be seated in one corner of the room with all the spectators facing you and have the room darkened. Remove one of the cardboards from your pocket, unpainted side toward spectators Now slowly turn the luminous side facing the spectator and the ghostly diamond pip will seem to materialize from nowhere. Move the cardboard around in a crazy fashion for several moments, then bring it to your mouth and hold one edge of the cardboard tightly between your teeth.

Produce the next two cardboards the same way except at one and the same time. Holding a cardboard in each hand move them slowing around, finally bringing one above and one below the cardboard held between the teeth so as to form the image of a three of diamonds. In the dark, this ghost of a three of diamonds will appear to be four or more feet in height.

Cause the diamond pips to vanish, one at a time, by simply turning the blank or black side towards the spectators, then secretly return the cardboards to your pocket.

Call for the lights up and have the spectator disclose the selected card, which in this instance will be the three of diamonds, the same as the ghost card.

Ghost Coin

An interesting effect would be to materialize the date on any BORROWED COIN. This effect can be accomplished by the use of two sets of cards bearing the numerals from zero to nine prepared as in the effect just described—Ghost Card.

Spirit Answer

A set of coardboards, each bearing a letter of the alphabet might easily be manipulated, one at a time, to spell out a spirit answer to a spectator's question.

Mysterious Dice

Here is a pocket trick with dice and a small black bag. Spectator thinks of a number, one to six, as shown on dice. He removes a die from the bag and places it upon the table with the number thought of on the top side so all present may note this number. The die is then replaced into the bag which in turn is handed to you, the performer. You reach into the bag, remove the die and immediately name the number thought of.

The bag is about four inches wide and six inches long, and is unprepared. A one inch die, however, is prepared. Secure a transparent die having white spots. Coat each of these spots with luminous paint.

Be certain there is a lighted reading lamp on the table used by the spectator. When the spectator places the die on the table, the luminous spots on the top side will be activated by the light from the reading lamp. In removing the die from the bag, you will readily see which side of the die shines brightly and the number on the shining side will be the original number thought of.

Mystery Magazine

This effect appeals to those with a taste for the literary. The spectator secretly selects a page in one of the more popular weekly publications such as the *Saturday Evening Post, Time, Life,* etc., and carefully notes the top line on the selected page and hands the magazine immediately to the magician only after closing it.

You are able to at once without hestitation, turn to the selected page and read aloud the top line.

The magazine is prepared by placing a minute spot of luminous paint on all the odd numbered pages.

For this test you select a spectator, for assistance, that is seated closely to a lighted reading lamp, saying: "In all magazines we have odd and even numbered pages. Will you kindly select aloud the odd or even pages." Regardless of the spectator's choice, you have eliminated half the pages in the magazine and now know whether the selected line will be on the odd or even numbered pages.

Request the spectator to mentally think of a number (odd or even as per his first selection) and between one and forty or between whatever number of pages are in the magazine being used. Hand the magazine to the spectator telling him to turn to the page corresponding with the number he mentally selected, and to secretly read the top line on the selected page. As the spectator reads, the light will activate luminous paint on this page.

Take the magazine from the spectator, and in the dark or by casting a shadow on the pages with your body, quickly glance through the pages and locate the spot of luminous paint that shines. Knowing beforehand that the selected page would either be odd or even numbered, you can easily pick out the correct page and read aloud the top line.

Mentally Speaking

A good phantom tray can easily be made by painting the inside bottom of a cheap ash tray with luminous paint. The tray is carried in the inside coat pocket.

Again select a spectator that is seated by a lighted lamp. Ask him to mentally picture some article he might happen to have on his person. A coin, key, knife, ring, match or the like for example. This done, turn your back towards the spectator and tell him to hold the mentally-pictured article in his hand.

Take the phantom tray from your inside coat pocket and hold it behind your back and directly in front of the spectator. At the same time tell the spectator to lay his article on the tray. Make a few remarks about the effect you are doing to give the luminous paint on the tray time to become activated by the light.

Request the spectator to remove the article and immediately put the tray back inside you coat pocket. When the shadow of your coat strikes the tray as you replace it in your inside coat pocket, you will notice a black silhouette of the article plainly visible and easily recognized.

Turn towards the spectator and dramatically name the article that a few moments previous had been placed on the phantom tray.

Evil-Eye

You can have lots of fun with an Evil-Eye when holding a dark room seance, especially when it floats in mid-air and over the spectator's heads.

With luminous paint make a large cat's eye on a piece of four inch square cardboard.

Secure a steel extension rule and have soldered to one end a short pin or small hook.

Before presenting this effect to an audience, expose the Evil-Eye to a bright electric light and place it, along with the prepared steel extension rule, in your pocket.

Have all the spectators in the room seated and facing you when the lights are extinguished. Bring the Evil-Eye into view and wave it around for several moments. Then fasten the cardboard to the pin or small hook on the ends of the extension rule. Slowly and noiselessly feed out the extension rule until you can float the Evil-Eye high in space and over the spectator's heads.

To make the Evil-Eye vanish, reverse the foregoing and when you have the gimmick safely in your pocket, call for the lights.

Spooky Lights

Spooky lights darting here and there around a dark room can always be relied upon for an eerie effect. Many and

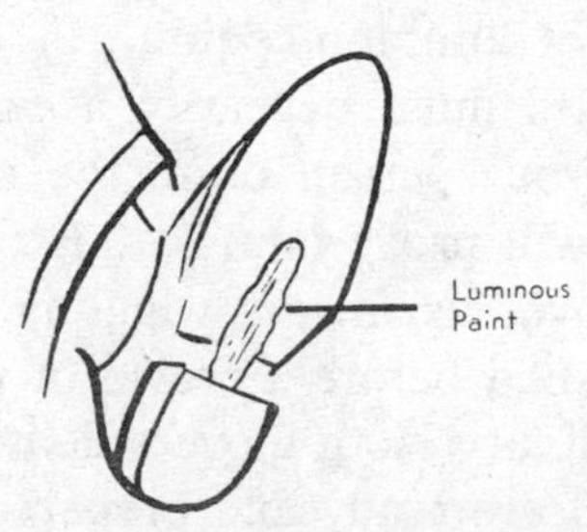

varied are the pieces of apparatus for accomplishing this effect, but undoubtedly the most simple method is to put a large spot of luminous paint on the bottom of each shoe, under the instep near the heel. In this position the paint is not disturbed by walking.

This effect is particularly good when those present are seated in a circle, in a dark room.

Before the demonstration you must expose the luminous spots to a bright light, then after being seated in the circle and all is in darkness just raise either foot (or both) as high as possible and wave them around. The spooky light becomes visible to the persons facing you. Bring them silently and lightly back to the floor and the light vanishes.

Probably, by this time, you have already realized the possibilities of presenting this effect under test conditions. Having your arms and hands securely bound or held by unknown spectators will not in any way effect your performing this experiment.

Apparitions

Apparitions of spirit articles are easly made and used. Just put a coat of luminous paint directly on the articles you desire for use and after exposing them to bright electric light they will shine in the dark.

Ordinary articles such as horns, bells, stars, etc., can be cut from cardboard and readily hidden about one's person. Paint the cardboard black on one side and give the opposite side a coat of luminous paint.

Luminous articles must be easily accessible in the dark, either hidden on your person or in the room. Fumbling or noise in the dark will prove fatal to this type of ghost effect.

A luminous glove, its fingers wiggling, will create a riot when suddenly flashed before a crowd of people in any dark place. Give an ordinary cloth glove a coat of luminous paint, and when dry, you are ready and prepared to go places, and do things.

Luminous paint is a valuable asset in commercial as well

as entertaining projects. A little applied to the dial of a watch or clock enables you to tell the time by night. You can coat push buttons, switch plates, match boxes, gas taps, house numbers, signs and innumerable other articles, making them visible in the dark for profit and pleasure.

Ghostly Faces

The late Sir Arthur Conan Doyle, it is said, claimed to have witnessed and recognized ghostly faces at several dark room seances which he attended.

You can duplicate this startling effect at a cost of but a few cents. Purchase cheap gauze masks, representing the types of persons you wish to have appear mysteriously in the dark. Give the gauze mask a coat of luminous paint, allowing a one-inch band around the edge of the mask to remain unpainted. By handling the prepared mask only on this unpainted portion in the dark, the fingers will not be seen.

When the mask is dry, remark the eyes, eyebrows, mouth, etc. with a soft black lead pencil. A man's, woman's and child's mask prepared as explained will remain flexible and may be hidden about your person until wanted.

Present this effect in a dark room with a little showmanship and you will be surprised the number of times the spectators' imagination will run wild, claiming they recognize the ghostly faces.

Ectoplasm

It is claimed that Ectoplasm is a spirit substance that comes from a medium's body, is visible in the dark, and after a short exposure returns from whence it came. It is said that should a person touch this Ectoplasm the medium would likely pass away.

This weird idea has been demonstrated by the use of many complicated devices, but here is a simple method whereby you can achieve the same effect.

Secure a piece of old time lace curtain, about sixteen inches long and five inches wide. Tear or cut this piece of lace into a pennant shape so that the edges will be rough and straggly. Coat the lace with luminous paint.

Before demonstrating, expose the lace to a bright light, then tuck it down between your collar and left side of the neck so that the narrow tip or end is easily accessible.

In the darkened room, grasp the tip of the luminous lace hidden behind your collar and slowly draw it from its place of concealment. Keep your head turned toward the left side and to a keen observer it will appear as if the luminous substance were coming from your mouth. Keep the lace near your body at all times.

To dematerialize the Ectoplasm, it is slowly rolled into a small ball and placed into a coat pocket. To really appreciate the value of this mysterious effect, try it several times before a large mirror in a dark room and you will undoubtedly be so well pleased that you will perform it over and over again.

Seeing Ghosts

The most intriguing and interesting of all luminous effects is the materialization of a ghost in the dark. Although ghosts are comparatively easy to construct and build, the success of a ghost materialization depends considerably on the showmanship and atmosphere created by the performer. In this type of seance, music and lights are a great asset.

One good method for making an inexpensive and useful ghost is to paint the figure of a ghost on a strip of black cloth with luminous paint. Then tack a black stick or batten across the top side of the cloth, just above the ghost's head.

When dry, expose the painted side to a bright electric light and then roll the cloth around the stick like a window blind.

To materialize this ghost, unroll the cloth with the blank (unpainted side) facing the spectators and hold it by the stick, slowly or instantly turn the luminous side towards the spectators. The speed of the materialization is easily governed. To dematerialize, just reverse the foregoing operation.

Another good method for creating a "ghost" is to use a wire frame that is shaped with a head and shoulders. To this frame is attached a luminous mask and several strips of luminous cheese cloth over the shoulders forming a flowing ghostly robe.

This type of ghost is concealed behind a large black screen or in a black bag until wanted, and is extremely suitable for ghost shows. It can be floated over the audience by simply attaching the ghost, which will be light in weight, to the end of a long black fishing pole. For a vanish just bring the ghost behind the black screen or if you desire a dissolving disappearance, gather it up slowly and put it into the black bag.

A large square of netting, painted with luminous paint, can be worn over your person and will make a very satisfactory appearing ghost for close-up work.

The Escaping Shadow

For this illusion you will need a box sixteen inches square and about fourteen inches deep with a strobalite treated screen on the front and an electrically equipped back panel.

Two armholes are cut in the sides four inches back of the front frame and six inches deep with a hinged flap to permit the arms being placed in the box. The wrists are secured with a Siberian Transport Chain or with "trick" handcuffs and placed in the box back of screen. Next the white light is turned on, then the red light, then the white light again and finally on dead contact.

The result, a green haze on the screen. The hands are then mysteriously released from the chains or handcuffs and the performer walks to the sides of the stage leaving the shadow on the screen and calls for the house lights. To the amazement of the audience he is totally free from the chains or handcuffs.

In showing the cabinet at first, do it with the front (strobalite treated) screen removed. Light the white, red and green lights respectively, then put them out and place the front screen on the cabinet.

Chapter **V**

Luminous Inks

Perhaps one of the least used and yet one of the most capable of all types of magic exhibiting a truly magical illusionary effect are luminous inks. These inks are capable of producing light from within its own chemical structure when in the dark. Simple enough to create, however, it is recommended that the intended user consider purchasing it already prepared from any one of the several sources of magical supplies.

For those that desire to try the preparation of their own luminous ink, there are provided formulas of the more general types in the following pages. Extreme caution, however, must be employed, for some of the chemicals used are not only highly poisonous, but are also capable of causing severe, painful burns that are very difficult to heal.

Luminous Ink

Dissolve:

Yellow phosphorous .... 1 part by weight
Cinnamon oil .... 8 parts by weight

in an oil bath over heat. Thoroughly mix this ink as the phosphorous dissolves with a glass stirring rod. In this way an ink that will shine in the dark may be prepared.

It cannot be cautioned too highly that PHOSPHOROUS is an explosive and should be handled only in accordance with the specified instructions found on the label of the container it comes in. Not to do so is liable to result in severe and painful burns.

This ink should be stored in a well-stoppered (glass), dark bottle.

Writing with this ink cannot be read in the light, but in a dark room it will appear as writing having been written with fire.

Fluorescent Inks

The science of crime detection laboratories has provided a still further type of luminous ink applicable to magic. Known as the fluorescent group, writing on white paper remains invisible until treated. However, upon exposure to ultra-violet light, Geissler tube or a high-frequency vacuum tube, a visible writing will appear.

For illustrative purposes several of such inks are listed as follows:

An aqueous solution of quinine sulfate will show up as blue writing when exposed accordingly.

An aqueous solution of uranium nitrate will show up as green writing when exposed as outlined in the foregoing.

An alcoholic solution of anthracene will mysteriously glow with a ghostly color when exposed as outlined in the foregoing.

An aqueous solution of eosin will mysteriously glow when exposed as outlined in the foregoing.

An alcoholic solution of naphthaline will also emit an eerie ghostly-like light when exposed as outlined in the foregoing.

Luminous Paste Inks

In addition to the previously discussed solutions, there are a few chemicals when mixed in the proportions of 1 part by weight chemicals with 100 parts by weight distilled water, used as ink, and then exposed to the invisible rays of light previously mentioned, will shine in the dark with mysterious, almost supernatural glow of light.

As follows:

Antimony oxide and calcium oxide	Yellow
Antimony trisulfate and calcium sulfide	Yellow
Manganous oxide and calcium oxide	Yellow
Manganous sulfide and calcium sulfide	Yellow
Bismuth sulfate and calcium sulfate	Red
Manganous carbonate and calcium carbonate	Red
Manganous phosphate and calcium phosphate	Red

Chapter VI

Daub

Magic Methods With "Daub"

(Author's note: With extreme thanks to Irv Weiner for the help in preparing this chapter, there is presented a few of the more illustrative types of daub and their application in magic. For those who are interested in the use of daub as an invisible accessory capable of producing myriads of miracles in the field of legerdemain, reference to *Manu Secret* by Irv Weiner is highly recommended.)

The preparation of daub has been reserved for and can be found in the chapter Chemical Tips for Better Tricks.

Examination of the magical market offers the selection of several types of daub. Marketed under various names, they all achieve a fundamental purpose, that of being a method of marking a card or several cards, which, under normal conditions and methods of handling, will remain invisible to the audience.

If ever the occasion should arise where the use of daub is required for "magical" use and none is available, one might use any of the following as substitutes—powdered graphite, carbon paper, water show colors, graphite from a lead pencil, rouge, etc.

Perhaps the most important factor in the use of daub is how to use it. The methods of application are numerous. Thus selection of a few of the more illustrative types ought to familiarize the performer with ideas on how to devise other methods.

From your favorite drug store, purchase a small (about one-quarter ounce size) ointment tin. It must be the metal type, flat in shape and round. Solder a safety pin to the bottom of the tin box in such a way as to be able to pin it to the bottom inside edge of your coat. In this way, you can "steal" a small amount of "daub" on your fingers and use it accordingly.

In the same way, for one using a piece of carbon paper, obtain a small piece of metal, solder a safety pin to one side and with rubber cement, glue the carbon paper to the other side. Use in the same manner as explained in the first method. This method presents a very effective way of providing a method for dispensing a daub substitute.

For powdered graphite, a metal plate as used in the carbon paper method with graphite rubbed onto a piece of paper in lieu of the carbon paper has proved highly effective.

The use of rouge is simple. Prepare the metal container with the pin soldered onto the back as explained before. Use in the same way as the others.

Even a button on your clothes is a convenient method for storing "daub." It offers an excellent method for being able to "steal" the daub as needed without detection by the audience.

For a final method, as written in *Dabbling with Daub* by Irv Weiner—"I have been told that Max Malini would carry both daub and wax wherever he went and he would secretly place a bit of each under the corner of the bar in a nightclub, then upon returning to the club at another time he was able to perform some of the most astonishing and apparently impromptu effects known."

Thus the other methods of possible concealment and the moves for obtaining daub are left to the imaginative minds of the magic profession. Once more, it is recommended that all magicians obtain a copy of *Dabbling With Daub* by Irv. Weiner.

Daub can be applied in several ways. All of these depend upon one thing, transfer of the daub from its source of supply to the object intended for reception.

To accomplish this is simple enough. The very actions you naturally follow in handling cards present many opportunities for the placing of daub.

As you fan the cards with their faces down, have a spectator select a card, push it out of the deck—half way—transfer the daub as you push the card out—let the spectator push the card back in. Immediately have the deck shuffled by the spectator and proceed further with your effect.

A second method would be to place some daub onto the palm of your hand, have spectator remove card from the deck and place face down on palm of your hand. Then place rest of deck on top, remove cards and shuffle. This has proven a very effective way to get daub on the face of the card.

Once again your own imagination will offer various methods of accomplishing miracles with daub. Use and guard its secret carefully.

Magic With Daub

A card prepared with daub can serve in either of two general ways—first, as a means of location of a card known to be in a specific position in the deck in relation to the daubed card—second, as the chosen card itself.

Inasmuch as magicians have been using the above methods with other means of applied deception to accomplish magic, only a few effects with the use of daub will be given as a means of illustration. The performer who desires to use daub in his magic will undoubtedly devise various effects capable of providing entertaining bewilderment to his audience.

Location

With a card somewhere in the middle of the deck already daubed, fan the deck (faces down) in an open fan. Allow a spectator to remove a card and note same and return to the deck. Pick up the cards after noting how far from the indicator (daubed) card the selected card is placed. Allow the deck to be cut several times.

It is easy to locate the selected card due to its position in the deck in reference to the locator card.

Imagination will invent various methods of final revelation.

Take A Card

In this effect the card selected by the spectator becomes the daubed card and is eventually found by the method of choice by the magician.

First, however, another way to secretly daub a card—this is believed to be an original method from the files of the authors of this book and is revealed here for the first time.

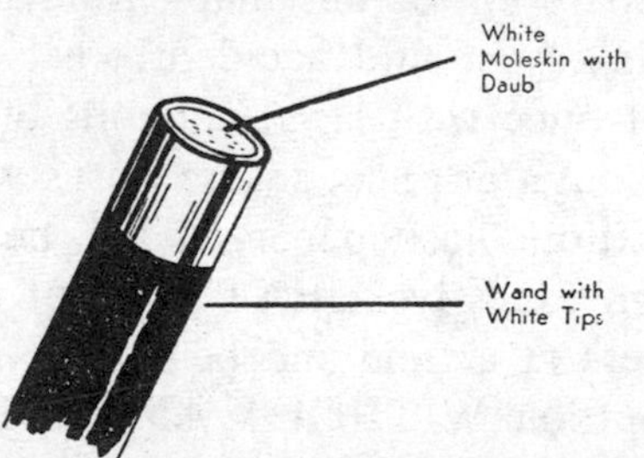

Take your wand and cut a piece of moleskin (obtainable from your drugstore) the exact shape of the end of your wand. Use the white kind to match the white wooden tips of the wand. With rubber cement, fasten the moleskin onto the end. The other side of the moleskin is coated (thinly) with daub.

To use this, as the spectator selects his card from amongst the deck, help him to remove the card by pushing it out with your wand. In this way, you secretly place daub on the card and continue from here to create miracles.

Chapter VII

Magic With Flash Paper

Flash Paper

A word of caution which cannot be stressed too strongly and which should be digested thoroughly, not once, but over and over again by the reader is indicated at this point.

Flash paper is an extremely delicate explosive which is liable to detonate at the slightest provocation during the process of manufacture.

Not only is this possibility ever prevailing, but the acids used in the preparation of this magical device are extremely corrosive and if any should accidently get on the hands or clothes, wash at once with large amounts of water. If this is not followed through at once, severe burns will result.

After preparation, flash paper MUST be stored in a dry place away from excessive HEAT. Do not make more than one or two sheets at a time and be sure to follow the directions for preparation WITHOUT ANY CHANGES.

First prepare solution of four parts concentrated sulfuric acid and five parts concentrated nitric acid. Place this mixture in a shallow glass dish. If any heat should develop upon mixing the acids, let the mixture stand until cool (room temperature).

Always pour the acids in small quantities when mixing and do not stand with exposed portions of the body too close. This will avoid possible splashing of the acids onto the clothes. Wear rubber gloves to protect the hands.

Soak white tissue paper about ten minutes in this acid solution. Remove the paper from the acid solution using a glass rod for this operation. Wash in running water for as long a time as it takes to remove all the acid. If the acid is not completely removed, the paper will be very unstable and an even more dangerous explosive will result.

After being certain that all the acid has been removed, dry in the open air away from all sparks or other sources of potential fire. Store single sheets between unprepared tissue paper or cardboard. DO NOT PLACE IN A CLOSED CONTAINER.

(Your magic dealer can supply you with flash paper which might be a safer way of obtaining it.)

Colored Flash Paper

To all intents and purposes it would appear that the flash paper you are using is the ordinary every-day variety. However, when ignited it is seen to burn in any one of six different colors.

The flash paper looks the same as ordinary paper and does not appear to be prepared in any way. To create this astounding miracle, first prepare different and separate solutions each containing one of the following chemicals:

For RED flash paper .... Strontium nitrate
For CRIMSON flash paper .... Strontium chloride
For YELLOW flash paper .... Sodium nitrate
For PALE GREEN flash paper .... Barium nitrate
For DEEP GREEN flash paper .... Copper chloride
For VIOLET flash paper .... Potassium nitrate

Use as much of the chemical that will dissolve in water to prepare the solutions required.

Colored Flash Paper

Another formula for prepared flash paper that will burn with different colors may be prepared by soaking in a solution of strontium chlorate (bright crimson), potassium nitrate (violet), copper chlorate (blue) and barium chlorate (green).

To prepare the required chlorates mix a warm solution of either strontium chloride, barium chloride, or copper chloride with an equal amount of a warm solution of potassium chlorate.

Pour off and save the clear liquid. This will represent the desired chlorate. Complete as discussed in the previous effect.

Dove Pan

Add small pieces of different color-burning flash paper to your chafing dish or dove pan prior to igniting same. This bit of added "magic" will create quite an impresison when performing this effect.

Chapter VIII

Fire Magic

There is little left in the realm of magic that in today's era one might call spectacular. Should a person overlook the field of fire, he might pass through a phase of magic common not only with the magician of today, of yesterday, but of the very essence of history itself. Fire has always created misticism in the minds of men from "day one that the very world was created."

Never before has any one single element of nature played such an unparalleled force destined to create and destroy history as has been done. Such is recorded in the pages of time.

History tells of Nero who fiddled while Rome burned, of the torches of life hidden in the bowels of the earth, of the sacrificial offerings of men to pagan idols who were thought to control the destiny of man with water, fire, etc. Fire was in truth a force of evil, because man who often created fire allowed himself to be destroyed by his creation.

Magic being an art embolizing mystery as did occur in ancient times, why not take the past into the present and use it in our magic of today.

Described in this chapter are illusionary effects, which if exhibited with a touch of showmanship will enhance any magical routine with mystery beyond the wildest hope of speculation.

Before delving into the true use of fire in magic, there are illustrated a few of the more novel and "common" uses of fire.

Since this is a book devoted to the use of chemicals in magic, the use of chemicals in fire hints of the use of pyrotechnics. Although not true magic in the strict sense of the "magical" word, the magician might find use for some of these in his portrayal of legerdemain.

As there is some danger involved in the practice of pyrotechny, certain fundamental rules should be observed:

(1) Unless you know for certain what you are doing, DON'T DO IT.

(2) Follow all instructions as stated. Do not change formulations given.

(3) Do not smoke while handling pyrotechnic chemicals.

(4) Always light pyrotechnic mixtures by external wicks or fuses or by other means thus explained. Do not use a match thrown into the mixture to ignite.

(5) Always mix and ignite outdoors.

(6) Avoid using large quantities.

(7) Although they are not in some cases, always treat explosives as capable of causing great damage if not careful. In this way habits of automatically exercising caution will be developed.

Blue Fire

Potassium chlorate	4 parts by weight
Copper sulfide	1 part by weight
Copper oxide	0.5 parts by weight
Mercurous chloride	2 parts by weight
Charcoal	0.5 parts by weight

Antimony sulfide	1 part by weight
Sulfur	2 parts by weight
Potassium nitrate	6 parts by weight

Sulfur	15 parts by weight
Potassium sulfate	15 parts by weight
Potassium nitrate	15 parts by weight
Potassium chlorate	28 parts by weight

Potassium chlorate	16 parts by weight
Mercurous chloride	8 parts by weight
Copper sulfate	10 parts by weight
Powdered shellac	6 parts by weight

Potassium chlorate	5 parts by weight
Copper chlorate	10 parts by weight
Alcohol	20 parts by weight
Water	50 parts by weight

Copper chlorate	50 parts by weight
Copper nitrate	25 parts by weight
Barium chlorate	13 parts by weight
Potassium chlorate	50 parts by weight
Alcohol	250 parts by weight
Water	500 parts by weight

Red Fire

Potassium chlorate	1 part by weight
Strontium nitrate	11 parts by weight
Sulfur	4 parts by weight
Charcoal	5 parts by weight

Strontium nitrate	8 parts by weight
Potassium chlorate	25 parts by weight
Mercurous chloride	8 parts by weight
Sulfur	5 parts by weight
Powdered shellac	2 parts by weight
Charcoal	2 parts by weight

Strontium nitrate	3 parts by weight
Powdered shellac	10 parts by weight
Potassium chlorate	5 parts by weight

Strontium nitrate	10 parts by weight
Potassium chlorate	5 parts by weight
Alcohol	15 parts by weight
Water	50 parts by weight

Green Fire

Potassium chlorate .... 3 parts by weight
Barium nitrate .... 8 parts by weight
Sulfur .... 3 parts by weight

Barium chlorate .... 2 parts by weight
Barium nitrate .... 5 parts by weight
Powdered shellac .... 20 parts by weight

Barium nitrate .... 36 parts by weight
Potassium chlorate .... 18 parts by weight
Sulfur .... 9 parts by weight
Powdered shellac .... 3 parts by weight
Mercurous chloride .... 6 parts by weight
Charcoal .... 3 parts by weight

Barium chlorate .... 10 parts by weight
Alcohol .... 20 parts by weight
Water .... 50 parts by weight

Barium nitrate .... 20 parts by weight
Potassium chlorate .... 20 parts by weight
Alcohol .... 40 parts by weight
Water .... 200 parts by weight

Yellow Fire

Potassium chlorate .... 18 parts by weight
Sodium oxalate .... 6 parts by weight
Sulfur .... 6 parts by weight
Powdered shellac .... 3 parts by weight

Sulfur .... 4 parts by weight
Sodium carbonate .... 6 parts by weight
Potassium chlorate .... 16 parts by weight

Sodium chlorate .... 10 parts by weight
Potassium oxalate .... 5 parts by weight
Alcohol .... 10 parts by weight
Water .... 50 parts by weight

White Fire

Gunpowder .... 15 parts by weight
Sulfur .... 22 parts by weight
Potassium nutrate .... 64 parts by weight

Wood charcoal	2 parts by weight
Sulfur	22 parts by weight
Potassium sulfide	76 parts by weight

Colored Flames

It is of interest just to note that certain chemicals when burned with a little alcohol will impart a color particular to that chemical to the flame.

Potassium nitrate	yellow
Sodium chlorate	yellow
Sodium chloride	yellow
Borax	green
Barium nitrate	green
Copper nitrate	green
Boric acid	green
Lithium chloride	purple
Calcium chloride	orange
Strontium nitrate	red
Potassium chlorate	violet

Water Starts A Fire

In performing this "magical miracle" of chemistry it is advisable to perform in an area with complete ventilation or even outside due to the dense fumes that develop.

Making sure that the iodine crystals and powdered aluminum used are absolutely dry, powder first the iodine and then mix with an equal amount of aluminum powder.

Place a small amount on the center of a piece of asbestos and add a drop of water on the top.

Without any delay, volumes of smoke colored purple-red will rise and a beautiful purple-red flame will develop.

Ice Starts A Fire

Like the previous experiment, this next is a vivid demonstration that not always can one depend upon ice or water to put out a fire.

Mix together 1 part of ammonium chloride with 4 parts of ammonium nitrate. On top of this mixture place 2 parts of zinc dust and finally, pour the entire mixture on a piece of melting ice. It will be seen to take fire and burn.

The Exploding Hammer

The exploding hammer is a very cute demonstration effect which is not very dangerous. Moderation in the quantities used hold the dangerous aspects of this effect to a minimum.

Prepare a mixture using equal parts of potassium chlorate and flowers of sulfur mixed with a small amount of glue. Mix in with this mixture a very small amount of powdered charcoal. Place a small amount of this mixture to the face of a hammer and allow to dry. When struck, a loud noise will result first.

Magic Sparks

Several chemicals will exhibit radiant colors when ignited. When dusted into an open flame (candle, fireplace, bonfire, etc.) colored sparks or flames will appear. Since these are flash effects, a magician contemplating a "chemical" magic routine has at his disposal a "flash opening effect."

Powdered aluminum	Silver sparks
Iron filings	Golden sparks
Aluminum and iron	Silver and golden sparks
Powdered lycopodium	Vivid flash of fire
Powdered flour	Vivid flash of fire

A mixture of barium chloride, strontium nitrate, calcium chloride, sodium chloride and potassium nitrate will exhibit an array of colors. Use equal parts (by weight) of each of the named chemicals in the preparation of this mixture.

Powdered magnesium metal thrown into an open flame will produce a vivid white flame. (Exercise extreme caution when using this effect, as the flash of fire is temporarily blinding due to the brightness.)

Flash Powders

A red flash powder may be prepared by mixing:

Powdered magnesium .... 2 parts by weight
Strontium nitrate .... 2 parts by weight

A green flash powder may be prepared by mixing:

Potassium nitrate .... 2 parts by weight
Powdered boric acid .... 2 parts by weight
Powdered magnesium .... 2 parts by weight
Powdered sulfur .... 2 parts by weight

A white flashlight powder is easily made by preparing a mixture of:

Potassium nitrate	2 parts by weight
Powdered magnesium	2 parts by weight

An extremely brilliant flash of light can be made by mixing together equal parts of aluminum powder and potassium dichromate.

A flash capable of giving off smoke as suitable for magical purposes can be made by mixing equal parts of black gunpowder and powdered magnesium.

Serpent's "Eggs"

Mix carefully, without rubbing or grinding.

Potassium nitrate	1 part by weight
Powdered sugar	1 part by weight
Powdered potassium dichromate	2 parts by weight

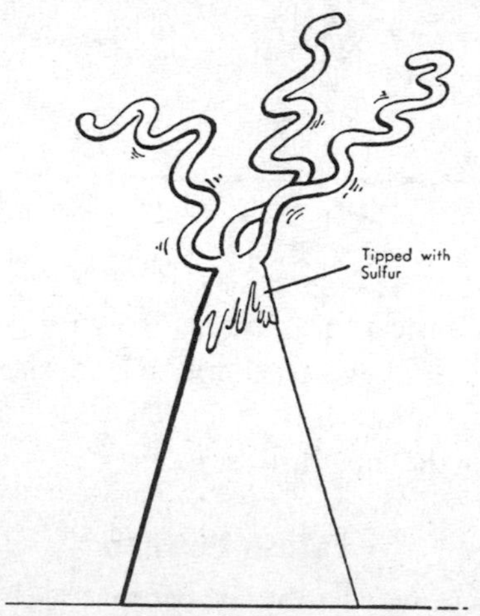

Add just enough mucilage of acacia to make a pliable and workable mass which can be molded into shape. Mold this mass into small cones about one-half inch high and set aside to dry.

When ignited by means of a lit match, they will produce an ash resembling a long continuous snake. The "snake" in reality looks as if there were more of it than the original.

Leamon, author of *A Handbook on Chemical Magic,* suggests tipping the cones with a bit of melted sulfur if you find them difficult to ignite.

Some More Snakes

Another chemical used to create "magical snakes" is mercury thiocyanate. Form into cones as in the previous effect and ignite. The voluminous ash formed resembles a snake and creates an illusion worthy of note.

Burning Ice

Mix nine volumes of denatured alcohol with one volume of a saturated solution of calcium acetate. A solidified mass will result resembling ice. To further promote the illusion of "ice," cut jagged chunks of this solid mass in order to complete the illusion.

This "ice" will burn when ignited. A suggested routine would be to hollow out a block of ice and fill the cavity with this solid "ice" prior to your show.

Allow some member of the audience to light the "ice" and as it burns you create a magical illusion that will certainly mystify all those present.

Smoke Screen

When exposed to air, titanium chloride exhibits strange properties. Because of these properties, it was and still is used by "sky writers" and has also found extensive use during the war for throwing a smoke screen.

Expose this chemical to air to create volumes of smoke. Its use in magic is unlimited. It might possibly find use as a screen to cover a production or vanish.

One idea might possibly be its use for a routine of smoke from the finger tips.

The Hot Gin

Pattering along the lines of magic being everywhere, on land, on sea, in the air, etc., you the magician explain that there is even magic in the decanter of water you are holding.

Water you continue to say has been used to put out fires, but what if situations were reversed and you exhibit some "ordinary" sawdust and proceed to pour some water onto it.

The sawdust bursts into flames!

Prepare a mixture of sodium peroxide and dry sawdust. It will be necessary to experiment to get the correct proportions. A few trials will fix the correct quantities to use.

If you want to, you can then demonstrate how the same water will extinguish the fire it has just created.

Fired Liquid

CAUTION: Handle this preparation as well as the chemicals making up the solution with extreme care. It is not only highly inflammable, but also is capable of causing severe burns.

Dissolve some yellow phosphorous in carbon disulfide. This preparation should always be made fresh as it is not stable. Store in a glass stoppered bottle as the carbon disulfide will attack and dissolve rubber.

Effects and routines using this liquid are described next.

Satan's Telegram

"A Western Union Messenger" comes down the aisle of the theatre paging the name of the magician. Walking on from the wings, the magician receives the telegram, opens it, and reads it. (Note: Any message suitable to the occasion.) As the magician finishes reading it, the telegram is seen to smoke and suddenly burst into flames.

Prepare some "Fired Liquid" as explained previously. The "Western Union Messenger" moistens the telegram just before coming down the aisle. The longer the walk, the lesser the amount of phosphorous or greater the amount of carbon disulfide should be used. The messenger places the telegram in the envelope, but does not seal it. Time is of importance in this effect in order that the telegram might be read before it starts to burn. Upon exposure to air, the carbon disulfide evaporates rapidly and the phosphorous remaining ignites the telegram which bursts into flames.

Think Fire

Saturate a piece of tissue paper with "Fired Liquid." Place on a glass tray and step aside. As the solvent (carbon disul-

fide) evaporates it will start to burn and will ignite the tissue paper.

You might patter how many people pray to many Gods, among which history relates of the God of Fire. Their prayers are a form of meditation and deep thought concentration. And if their prayers are to be answered, if they prayed deeply and strongly, a sign or symbol representing an answer will appear.

Judging your patter until the fire appears gives you a "thought-time controlled" chemical reaction of causing fire to appear because "fire was thought of." A bit dramatic, but after all, isn't all successful magic based slightly upon dramatic presentation?

An Innovation

Why not use "magic" in your dove pan? Place the bits of tissue paper wet with the "Fired liquid" in the dove pan and patter about heat being necessary to bake a cake. Being a magician you merely point your wand at the dove pan and there is fire. Same principle as described in "Think Fire." Cover pan and make your production.

The Fire Bowl

Prepare a brass bowl by placing in it 10 cc. ether and a piece of potassium metal the size of a pea. Do not allow it to stand too long as the ether is highly volatile. When ready to cause flames to appear, pour some water into the bowl. When water hits the potassium metal, a reaction takes place which will set the ether burning while it floats on top of the water.

It is important that you read all the directions concerning the handling of both the ether and potassium metal.

Self-Lighting Candles

Prepare two candles by hollowing out a shallow ring around the wicks.

Place a few crystals of chromic acid in one candle and a few drops of absolute methyl alcohol in the other.

When the candle with alcohol is touched to the candle with the chromic acid, they will take fire and the wicks will ignite.

CAUTION: The candles must be prepared shortly before using, since the alcohol evaporates rapidly. The chromic acid is very deliquescent, that is, it takes up moisture from the air. If chromic acid is allowed to become wet, the effect will not work.

Be careful in handling the chromic acid as it will cause ugly burns on contact with the skin.

A Magic Picture

Make a line drawing on white paper, preferably a high grade white tissue, with a saturated solution of potassium nitrate. A small size clean camel's hair brush is best to use in applying the solution. Mount the picture in a frame, using thumb tacks.

Start by applying a spark, the lit end of a cigarette or cigar, or similar means of providing a spark to the beginning of the drawing. If you have used care in seeing to it that the line is continuous, the glowing spark will follow the line and burn out a picture. There will be no flame. The line will burn as a fuse burns.

It is possible to obtain very fine effects with this principle of chemical magic. A double frame can be made using black cardboard, cloth, or metal set about an inch behind the tissue. When the picture has burned around, the center will fall out—leaving the appearance of a silhouette. If heavy cardboard is used, the picture will appear as a black line drawing. Combinations of this effect with forced cards, or in a mind reading act are startling.

To increase the vividness and visibility of the flame in this effect add a little strontium nitrate or sodium chloride to the potassium nitrate.

This effect should find considerable use in spook shows.

Lighting A Cigarette With Ice

When one thinks of water to start a fire, he is often laughed at, for everyone knows that water is used to put out a fire—or is it?

For this demonstration we need some ice water in a frozen state and a cigarette with a piece of potassium metal, the size of the head of a pin inserted in end.

Touch the end of the cigarette with the ice making sure it comes in contact with the potassium metal.

The cigarette will immediately catch fire.

Candles Which Go Out

Cut a candle cleanly into two pieces. Remove about one-half inch of wick from one end of both pieces. Warm these two ends and stick the candle together again.

When the candle is lit by some unsuspecting individual, it will suddenly go out without warning, the time depending upon the exact position in which you divided the candle.

Chapter IX

Fire Eating

There is little left in the realm of magic that in today's era one might call spectacular. Should a person overlook the sensationalism of "fire-eating" which offers the magician perhaps one more item for an audience to long talk about, he might well decide to continue doing "routine" magic. If he desires to sensationalize his act, he might do well to include one or more "fire-eating" illusionary effects.

Since this is a broad subject and many excellent books have been written on the subject, there are illustrated and explained only a few of the more common effects. For those desiring more knowledge on this subject, your magic dealer can supply you with a choice of books.

In fire-eating there are several cardinal rules to observe. These are not only important to ensure success, but that also the danger element will be eliminated.

(1) The mouth should thoroughly be coated with saliva before placing lighted materials in the mouth.

(2) *Always* exhale the breath continuously through the mouth with a force as determined by the amount of heat placed in the mouth.

(3) If the lighted material seems to burn the mouth, do not attempt to rid it from the mouth quickly. Instead, shut the mouth tight and breathe through the nose. This will cause the fire to go out.

(4) Above all—always observe carefulness. Do not attempt a new effect until you are satisfied that you

can do it with the utmost perfection. Remember that a "show-off" usually is never around long enough to show his magic off.

AND FINALLY—MOST IMPORTANT RULE OF ALL......

(5) Unless you know what you are doing, don't do it.

Blowing A Paper On Fire

Prepare some "Fired Liquid" as explained in the chapter Fire Magic. Saturate a small piece of tissue paper with this liquid. Do not get any on the hands, as severe burns will result. With the wet paper on a glass dish, gently blow on the tissue paper and as the solvent evaporates, the phosphorous will ignite and the paper will burn.

Flames From The Mouth

I

A piece of brimstone wet with gasoline, lighted and placed in the mouth will cause vigorous flames to come forth by exhaling very gently through the mouth. A candle recently lit and put out may be ignited again.

II

Chew some cotton unprepared by chewing it in a small wad, remove secretly while placing a second wad into the mouth. Repeat this procedure again, except this time secretly have a piece of lit punk inside the cotton. When in your mouth, blow (exhale) through the mouth and fire and sparks will appear.

"Eating" Fire

I

Wet several small sponges in alcohol and set them on fire. Open the mouth and keep the head thrown back. Drop a blazing sponge in the mouth and close it quickly. This tends to extinguish the flame. The illusion is perfect. You have just finished a meal of fire.

II

Ignite one-half inch squares of camphor, toss them from hand-to-hand one at a time, throw them up in the air and catch in the mouth, exhaling through the mouth as you do. Do this three times. Wipe mouth with a napkin removing the camphor at the same time. Practice this carefully as the illusion is perfect with the proper presentation and misdirection.

III

Prepare a saturated solution of potassium nitrate in water. Soak a thick soft card in it for at least twenty-four hours. Take out and dry. Cut into pieces about one-fourth inch long by about one-half inch wide. Light one of these pieces and roll it up loosely in some cotton, making a ball about one inch in diameter. Place in the mouth and inhale through the nose and exhale through the mouth. Doing this will cause showers of sparks to fly out. To further enhance this, pretend to eat a ball of cotton every time you get ready to blow more sparks out of the mouth.

IV

As with the sponge trick in this section, pieces of bananas dipped in alcohol—lit—placed in the mouth will go out and may actually be consumed. This creates the "perfect" illusion of eating fire.

The same may be done with other fruits (apples, raisins, pears, etc.).

V

Cut a piece of apple to look like a candle or use a piece of banana. Insert a short piece of cotton string for the wick and light it. As in the effect just described, the flames go out when placed in the mouth, the candle (apple, banana, etc.) is consumed and you create a reputation for yourself.

Chapter X

Magic With Slates

The authors extend their thanks to Burling "Volta" Hull for permission to reprint from *The Invisible Hand Writes* his very excellent and mystifying routines using the principles of visible slate magic.

The effect in brief—a slate (or several) is shown blank, a question to the "spirit world" is written thereon—visibly washed off with water and allowed to dry. Slowly the answer is seen to appear as if being written by magic. In many cases the slate at this point can be passed for examination by the audience.

Mr. Hull has written patter effective for this type of routine and is included here to illustrate an idea. It should, however, be understood by all that digest these pages, that to be effective in presentation, orginality is of prime importance; and the performer should adapt patter suitable to his own routine and personality.

From Mr. Hull's own repertoire, this is his patter and presentation:

"Wheeling about and pointing dramatically to the blank slate—'I will now call upon my Spirit Control, dwelling on the other side of the great divide, to make his (or her) presence known to us, by writing a message on the slate. A message from the Spirit World. Please write a message answering our question (or questions). Ready—come on—I call upon you to respond—everyone here, please concentrate and help us in this effort—lend me the dynamic power of your personal thoughts to will the spirits to respond—come—begin—now—I command you to commence to write.' The above wording is so arranged that at any point when the writing starts to appear, you can stop your patter and the writing will seem to appear at exactly that second. (The thing to consider is that the temperature of the room or if there is a slight breeze present the appearance of the writing is speeded up. In extremely cold or damp weather, the writing slows up.)

"As the first letter of the writing appears (it usually appears in the left upper corner of the slate) stop your running patter, and say, 'Ah! There it is! The spirits are slowly tracing some words for us on the blank slate, in response to our command. Some of you in the front rows where I am standing can see a sort of vapory transparent hand, like a faint cloud of smoke gliding over the slate tracing the letters.'

"A number of persons have nodded their heads and later assured me that they did actually detect the phenomena that I described. Try it and see, the illusion is almost perfect."

The Scientific Secret

The scientific secret of this remarkable effect lies in the use of a specially prepared chalk. Messages written with this chalk and subsequently washed off with a specially prepared liquid (looks like water) has the property of flashing up, after a few seconds, as snow-white chalk writing. In fact, close inspection with ordinary chalk, the writing appears exactly identical.

The important point is, that the appearance can be controlled by the performer so the writing does not all develop at once, but the first word of a sentence may be caused to appear, then the next word, etc. Line by line the message materializes.

Mr. Hull has perfected a highly improved chalk and "wash-off" liquid available for sale to the magic fraternity. There are several other methods for accomplishing this effect and they will be discussed later. From actual experience, however, it is highly recommended that the performer using any of the foregoing routines purchase and use the improved materials already mentioned.

"The Spirits Know All" or a Living and Dead Combination

This routine is an impressive and decidedly different routine of a spiritualistic interlude. It might well be described as being uncannily mysterious.

Have a person think of the name of some dead individual, preferably one who is not too prominent, one, as the performer explains, not recognizable by him if he were to see it.

To this spectator give several slips of paper along with the instructions to write the name of the person so selected and also the names of several different living persons, one on each of the remaining slips.

Collect the slips and mix them at random and lay them to one side. Pick up the slate and write the names in chalk on the slate. Call attention to the fact that it is impossible for you at this time to know who the dead person is. The spectator will agree you are correct and you conclude the effect by washing the slate clean, remarking that he (the spectator) is wrong for the spirits present in the room know. As the magician patters, slowly, to the amazement of all, writing slowly appears on the slate revealing the name of the dead person.

The secret and principle back of this effect is derived from two different presentations, the first being the appearance of the name on the same side of the slate where the names were written, the second with the name appearing on the back or opposite side of the slate.

Still another version is to clean off the names, place the slate again upright on the stand and walk into the audience. The name will appear in large giant writing covering the entire side of the slate.

U. F. Grant has furnished the principles and method for accomplishing the first method. The performer merely has two small pieces of chalk about one inch long, which he picks up together from his table. With the dry chalk, (the plain chalk) he writes the names of the living persons. When writing the name of the dead person, he shifts the chalks in his hand and uses the chemically prepared chalk.

This should be second of the five names so it will allow a moment or two for the chalk to dry and become visible and match the other chalks. By the time you have written all five names, and carried the slips back to the spectator in the audience, the dead name will be dry as the others. Show the slate to the audience.

It is obvious that you merely have to patter a minute calling to the fairness of the experiment. Have the spectator write the dead name on a page of the pad you left with him. Fold it a couple of times, write his name on it and hand it to another spectator for safe keeping. This is so the audience will be able to identify the name. It is really to give you a few more moments for thorough drying. Cleaning off the slate with the specially prepared "wash-off" liquid causes the name to flash up alone. This takes about 15 to 20 seconds to occur.

For a second method, Mr. Hull has created this method of performing the above-mentioned effect.

The performer steps off for a moment to get the slate while the spectator is writing the name on the pad. He quickly writes with the special chalk the name of the dead person, writing diagonally in large letters from corner to corner. As the chalk is moist, it will make no noise so it cannot be heard. Don't stop to write carefully, write quickly for the spirits would very likely write it in a scribbly hand if they did it themselves.

Place your neat fitting silicate flap over the slate and come out. Write the names on the flap, in this case all five with plain ordinary chalk. Then as described, erase the names. While commenting to the audience on the fact that you have no way of knowing which of the names is that of the deceased individual, lay the slate down for an instant while you pour a little more liquid into your glass cup, pick up the slate (leaving the flap on the table) and holding the written side toward backstage (away from the eyes of the audience resume washing off that side as if it had not been completely erased in the first washing. This renders the "dead" name invisible before you turn the slate around and fit it into the stand and set it upright on the table.

It only remains for you to walk away, point at the slate, the dead name flashes up in giant writing on the same side, as the others were written on. This is a very dramatic spirit effect.

A Daring Volta Method

(Authors' note: There are usually a number of ways to do the same "magic" effect. In some instances there is always a way which will far excel the others. In the above-mentioned effect Mr. Hull has achieved such a routine. It takes a person with self-confidence and showmanship to utilize this method; and with practice, an effect can be created destined to fool even those in the "know-how.")

In Mr. Hull's own words, here is his method:

"Before the audience is seated, the performer remembers the name of a person who will be present in the audience, such as the president or a prominent member, writes in diagonally across the flap of his slate. He places it face down on his table. He proceeds with his performance in the regular manner until he comes to this trick and then—

"While pretending to write the name of a spectator on the slate (the face of which is toward him and away from the audience) he really writes (with the chemically prepared chalk) the name of the "dead" person. The audience cannot see what he is writing and assumes it is the name of the spectator whose "spirit" relative you will call upon.

He lays the slate down over the flap, while he (performer) carries the slips of paper down into the audience, etc., as described about.

"When he returns to the table, picks up the slate with the flap under it and carelessly turns the slate around exposing the name of the spectator.

"You can see the cleverness of this method. It has mystified magicians knowing the general idea, but could not understand how the performer could get the name of the 'dead' individual written on the back of the slate without leaving the stage. It is obvious he could not know before the performance what names were going to be selected.

"Now erase the names from the front of the slate, lay the slate down a second while you pour out more liquid (getting rid of the flap). Clean off other side (keeping the back out of view) before turning it around. Place the seemingly blank slate on the stand and say that you are going to call on the 'spirits' to write in the name of the dead person on the slate. The name then appears diagonally from corner to corner on the slate."

Dead or Alive Discovery

In case you are puzzled as to how the performer knows which is the "dead" name, we will explain this method as used by "Spirit Mediums" and which is a very effective method. Guard its secret carefully.

Use a soft pencil (No. 1 or 2 grade) and fairly rough paper slips. Sharpen the pencil to an extremely fine point. Have the first name written to be the "dead" one, and this will be written with very thin lines at the beginning of the name because the point is so sharp at the beginning of the writing, but as the lead is soft, and the paper is rough, it will quickly wear down the keenness of the point. All the other names, the living ones, will be written by broader pencil lines as the point broadens down with use. This is an old "Spirit Medium" stunt which has been "lifted" from the Mediumistic professions. As the strips are laid out in front of you, while you are copying them down on the slate, it is much easier to pick out the "dead" name that appears before you amongst the entire group of slips.

A Number Test

Along the top of the slate are written a series of numbers in large letters. The performer hands the slate and a piece of chalk to the spectator. He asks them to merely think of any three numbers and keep them in mind.

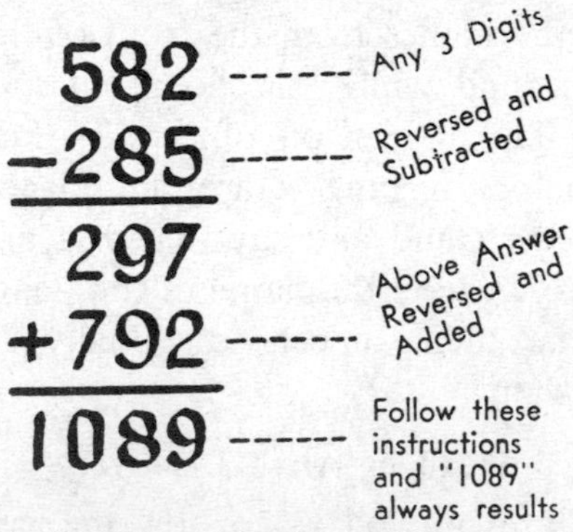

582 ------ Any 3 Digits
−285 ------ Reversed and Subtracted
297
+792 ------ Above Answer Reversed and Added
1089 ------ Follow these instructions and "1089" always results

"I would like to have some gentleman or some lady in the audience assist me by thinking, yes merely thinking of any three digits or single figures, like 5-8-2 or 9-1-3. Any figures whatever will do. You, sir, thank you. Now be sure you don't change your mind about those numbers!

"Here, sir, is a slate, and I want you to write down the figures which you are thinking of so that you will not forget them, and you cannot change your mind. There is a row of figures on the top of this slate used in a previous experiment, but as they have nothing to do with our present effect in magic, just disregard them.

"You have written the numbers, thank you. Now reverse the figures writing the last one first, and the first one last, etc."

(Here the performer illustrates by writing a set of three figures and reversing them.)

"But, just a minute, if your figures when reversed are larger than these original three figures, then place the larger figure above the original. If it is smaller, place it below.

Now subtract the smaller row or line of figures from the larger. Finished—good! Now reverse this answer, then add the two figures up. Ready? Have you the answer? Good!

Now for the climax. I am going to take this slate from you (holding it high above head and walking back to the platform so audience can see that no exchanges are made) and clean it off with this moist sponge (doing so with slate held in front so performer cannot see the numbers written by the spectator), so I cannot see the answer. Once again I will clean it off so that the slate is entirely blank. I will place the slate down into the audience and call upon the "spirits" to write and tell us the answer which you arrived at in your own example, and from the three figures which you selected mentally of your own free choice. Impossible—well perhaps, but we shall see."

Slowly the answer appears along the top of the slate, "1089."

"Was that your answer, Sir? Thank you, and I personally thank you and the 'spirits' for their most generous assistance. They did read your mind and discovered the numbers which you have thought of mentally, worked the example for us and wrote out the same answer, which you personally arrived at, from your own numbers."

The Secret: Across the slate near the top write with the

specially prepared chemical chalk, 1 0 8 9, but well spaced out so you will have room to write other figures in between. Now between these figures fill in with ordinary chalk other figures. Also add an extra figure before the beginning and after the end of the line.

Another method might be (according to "Volta") is to trace over the numbers 1 0 8 9 with ordinary chalk changing them into other numbers. For example, the 1 can easily be changed to a 7, the 9 can be made to look like an 8, etc. A novel idea would be to interchange the writing using colored chalk.

(Authors' note: The use of colored chalk has been proven more misleading than just using the white. This has been seen from actual performing experience.)

Why always the same numbers 1 0 8 9? According to the useful magical principle, any series of 3 single digit numbers reversed and subtraction of the now lower number from the higher, then reversing the answer and adding it back upon itself will always result in 1 0 8 9 as the answer.

When the slate is washed with the specially prepared wash-off liquid, the numbers written with "ordinary" chalk wash off and the prepared chalk numbers flash into view.

An unlimited magical mystery can be made by the use of a few original additive ideas.

The "Spirits" Correct A Mistake

The performer during the course of his program has several cards selected (say, 3 or 4) and then proceeds to discover them, or uses these selections in any experiments in his regular routine.

The names of the other cards are given correctly or discovered by the performer, all except one card, say the five of hearts. The performer seems very crestfallen and turns for help to the "spirits." Picking up the slate, he exhibits it covered with writing which makes no sense. By way of explanation, he states that it is a code message given to him by the spirits from the great beyond to enable him to correct any magical revelations which are wrong.

Showing the slates with the "code message," he cleans it off with a moist cotton or sponge, stands it upright on the table and then walks down to the audience.

Slowly as the spectator concentrates upon his card, the name mysteriously flashes on in full view, this time it being the correct one.

As in the foregoing effect the letters (five of hearts) can be changed into other letters. The "F" can be made to look like an "R," the "I" into a "T," the "F" into a "P," the "E" into a "B," the "R" into a "B," etc. Then fill in with additional letters in between and at the ends of the message.

A Few Remarks

Mr. Burling "Volta" Hull has advanced quite a few original ideas into this type of magic with slates and has prepared a few remarks considered highly important by the authors from the point of view of doing "good" visible slate writing and being able to devise several original routines.

With his permission, there is reproduced these "remarks." Digest them thoroughly, for to understand them is to understand how to accomplish "good" magic via the aid of visible slate writing.

(1) The message decided upon should be brief as possible, because the fewer the words, the larger the "spirit writing" may be made.

(2) The message must be written very lightly, with very light pressure on the silicate slate. Then it must be allowed to dry for about fifteen to twenty minutes in cool climates, a little less will do in hot dry weather.

(3) Only very "absorbent" silicate slates may be used for good effects. Ordinary stone school slate will not work. Neither will hard composition blackboard material made over masonite, plywood, composition board, etc. These will not ab-

sorb and hold sufficient wash-off liquid to render the spirit writing invisible at a short distance, nor will it allow the desired effect of the message gradually appearing.

(4) There are three procedures that may be used according to which effect you desire to produce.

a. *Blank Slate:* To bring out a blank slate to be shown on both sides. The first method is to make invisible the chalk writing (chemical) by flowing high test naphtha on the slate. Let it stand for a few seconds and blot it off lightly with cotton. Do not rub slate hard. Later when you are ready to perform the effect, wash the slate with the special wash-off liquid and since this is fast drying, the snow white (chemical) chalk writing will begin to again become visible.

b. *Flap:* Prepared message on slate covered with flap. Write with ordinary chalk on flap and then when ready to wash the slate, remove flap in your favorite conventional manner. With the message now towards you, wash it off with special liquid and then turn the slate around to face the audience. The message appears as the wash-off liquid evaporates.

c. *Disguised Message:* Wherein the message is on the slate in full view of the audience, except it has been changed by making the numbers and/or letters comprising the intended message appear as others by writing over them with either white or colored (unprepared) chalk.

(5) Gradual appearance of a message may be created by writing the message from the top left corner diagonally to the opposite corner. Wash the message off in the same manner and as the message appears it will seem to materialize letter by letter almost as if an invisible hand were in the process of writing the name. A little experimentation is necessary to achieve the desired effect.

A Mind Reading Number Divination

Performer passes out several cards and pencils, asking the spectators to write single digits or figures upon them.

These are passed along up to him. On a large slate he writes down the numbers furnished by the spectators.

A spectator is then asked to mentally select one of the rows, either vertical or horizontal and then to write those figures down on another card which he (the performer) hands him. As soon as that is completed, the performer cleans off the slate and stands it upright in full view.

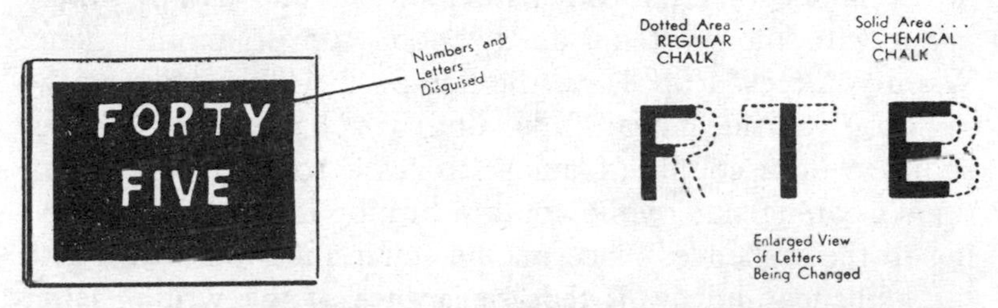

"I will write the spirit world a message" (performer writes). "Please give us the total of the figures which the gentleman in the audience has just selected. I guess the spirits have received our message by this time so I will clean it off."

He lays the slate down (the now cleaned—???—side) on top of another slate, while he pours a little more liquid into the cup as he seems to need more of the liquid. Picking up the slate, he cleans off the back before turning it around. Standing the slate in the frame, the invisible spirits are seen to be writing the message.

In the upper left corner of the blank slate there gradually appears a white line which curls around and forms a large letter "F." Then the writing continues diagonally across the slate to cause the following letters to appear "O," "R," "T," "Y," then "F," "I," "V," "E"—until the audience can read the words "Forty-Five." The slate is then passed out into the audience, the writing examined, found to be "pure" white chalk. It is compared to the total arrived at by the spectator and found to be correct.

This feat may be accomplished with or without a "flap."

If the diagonal writing is to appear, starting at the upper left corner, with a single letter and continuing diagonally across the slate, letter by letter—either of the following procedures may be used.

With a Flap: Before the performance, write the words "Forty-Five" in giant letters, diagonally across the slate using the specially prepared chemical chalk. Cover it with the flap. When laying the slate down to pour out the liquid, simply lay it with the written side down on top of another slate. When you pick it up, leave the flap on the table, keeping the "Forty-Five" side toward you, until you have wiped the wet sponge with a couple of quick strokes across the slate. Continue cleaning slate while you are turning it around and talking to the audience. Place on the stand and walk down into the audience and await the appearance of the writing letter by letter (appearing gradually).

Without a Flap: This is a cute wrinkle and it's novelty will appeal to the clever magician. On the back of the slate, write the "Forty-Five" diagonally across slate with the chemical chalk. Then with ordinary chalk fill in the same numbers which you are later going to write on the front of the slate (you will see later why you will know what these numbers are going to be before the performance). It is better to use a smaller slate not over ten inches high for this method.

After the spectator has selected his row of figures, and written them down, start to clean off the slate with an only slightly moist sponge, wiping sponge diagonally across the slate just where the "Forty-Five" would be if it were written on that side.

Turn back to your table to get more liquid on the sponge and under cover of your body, give the slate a half turn. Immediately placing the now thoroughly wet sponge on the slate, make a quick broad stroke directly over the "Forty-Five" and continue scrubbing off the figures below this while bringing the slate around into view of the audience. To spectators it appears that you are merely continuing to wash off the table of figures and nothing else.

Of course the table of figures written in ordinary chalk is completely washed off, while the "Forty-Five" written in the special chemical chalk, will return in a few seconds, after you have walked into the audience. Starting with the first letter, it will appear letter by letter as if written by the hand of invisibility.

Selection of Number: How does the performer know the total of the row of figures mentally selected secretly by the spectator? When he gathered the cards bearing all sorts of figures, he said, "There are certainly plenty of them," and started writing them in. But he really wrote in his own digits to form a magic square. You know that in a magic square every row whether vertical or lateral (even diagonal) will add up to exactly the same total. If the performer cannot remember his magic square, he just has it written down on one of the cards on his table which he did not distribute to the audience, he really copies in his magic square figures.

When he gets them all written down, he says, "Well I guess that's enough," and then stops. Not one person in fifty in the audience knows what a magic square is or will connect with the lot of figures because he thinks the numbers from the audience are being used. (Use any other magic square instead of the square of Forty-Five—if you prefer.)

And that is why the performer can have a duplicate set of numbers written on the back of the slate to match those he is going to write on the front, in case that part of the operation has been puzzling you up to now!

Another "No Flap" Method

Here is another cute method of performing the feat, eliminating any of the operations required in the former. The performer has a second slate standing upright on table, bearing some indifferent letters. We will call this slate No. 2. On the No. 1 slate which is blank, he writes down the digits provided by the audience. These are actually the magic square figures as in the foregoing effect. After one row is selected, and copied down by the spectator on his card, performer asks which row it was. On being informed, he wipes out the other rows leaving only the selected row, and stands the slate upright in full view.

He then makes his announcement, about calling on the "spirits" to add up the figures, and tell what the total of the selected row will be. He picks up the second or No. 2 slate and washes it off with his wet sponge, squeezing sponge so as to get lots of liquid on the slate, turns it upright in stand and walks quickly into the audience. At the word of "command" the "spirits" start to write out the total—first the word "Forty" (a slight pause), and then "Five" flashes up underneath.

Spirits from China

The use of the Chinese language in characters offers a very good opportunity of creating a very mystifying effect using Chinese characters.

With the principle of disguising writing by changing to look like other letters, Chinese writing will be found very easy to change to other letters (characters).

The use of Chinese proverbs might be one suggestion. If your favorite Chinese laundryman is particularly bright, have him select a proverb full of meaning suitable for the performance that evening.

With your selected proverb written on the slate, using the special chemical chalk, place a flap on the slate. Have some member of the audience write a question on the slate (actually on the flap). Take slate, and as you pick up a sponge to wet with the wash-off liquid, you lay the slate down on the table leaving the flap there. Pick up slate (minus flap) and wash off (audience thinks it is the question) with the liquid. Place slate in stand and answer soon appears.

The performer steps back on the platform and says, "Naturally I handn't thought of the fact that our Chinese Mind Reader would naturally write his answer in the language familiar to him—Chinese. It is a little unexpected, but fortunately I have a pretty fair knowledge of Chinese gained while associating with a famous Chinese Magician, Ching Lee Foo. So I will translate it for you. It says, (here performer gives his own made up answer).

Suffice to Say

Be it the Dunninger slate (often called the Al Baker Slate) in doing the conventional effect where rows of single digits, each row containing four numbers, are written down on the slate by four different persons in the audience and then added by the spectators, the performer then predicts the answer which has previously been written by him prior to the doing of this effect—the climax is terrific.

Chemical (visible) slate writing offers a new twist to this. The agile mind after thoroughly digesting this chapter can and will be able to devise various methods of approach to the final revelation of the answer.

A new twist on a grand old trick which should make all those that say, "I saw that one last time," sit up and take notice.

Colored Chalk

There is available on the market colored chemically prepared chalk. Your dealer can supply you with this item.

A possible use for this type of chalk can well fit in a children's magic show.

For Children Only

The author devised this routine for use at children's parties of the grammar school age.

On a slate draw the usual comic representation of "teacher" shown in so many cartoons. Cover this colored chalk (chemical chalk) drawing with a loose fitting flap. When performing the effect patter along the lines that you should never draw a picture of the teacher on the blackboard especially one like this (draw a representation of "teacher"). Under it draw the word "teacher." Use colored chalk (regular type). Continue on by commenting that when young you were caught doing such a thing and the teacher came in the room and caught you. You replied to her by saying nothing and hurriedly erased the board with the wet sponge you had.

As you pick up the sponge, place the slate down leaving the flap on your table. Pick up slate with colored chalk (chemical chalk) drawing facing you away from audience. Wash off with liquid and turn to show to audience, saying that it is wrong to draw a picture of teacher and as you continue to patter, pay no attention to the slate. The expressions of laughter on the faces of the audience as the picture "slowly develops" into view is reward enough for a good routine.

Additional Ideas

U. F. Grant has created several ideas for the use of visible slate magic in combination with pictures and magical effects. Mr. Hull has consented to the reproduction of these ideas from his book *The Invisible Hand Writes*. Thanks to both Burling "Volta" Hull and U. F. Grant.

One of the Three Little Pigs

Tell the story about the small boy who brought his slate home from school and asked his father to draw one of the three little pigs for him. His father not being much of an artist, drew a picture of a square pig. Upon showing it, his son remarked, "That doesn't look much like a pig. Pigs are round. Can't you make it look round?"

Then wash off the picture of a square pig with the special liquid and continue by saying, "Why certainly I can." So he washed the slate off with some magic liquid he had and sure enough, the pig appears "round looking."

The Rising Card

On the flap draw a sketch of a deck of cards in a glass goblet. Then have a card selected from a regular deck (forcing the card). Now wash slate off and as the goblet and deck reappear, a card is now half way out of the deck as if it had ridden out. It is a duplicate of the selected one. Best to force an ace, as that is the easiest to draw on the slate with chalk. Draw in colors or in white chalk.

Novelty Balloon Stunt

Show a blank slate and draw a picture of a deflated balloon. Say, "And then along came a boy. You know what happens when boy meets balloon!" Place slate down when you pick up your "water," etc., and wash off, leaving flap with drawing on table. Wash off slate, and the balloon reappears all blown up, filling the whole slate, with a small body down in one corner. It gives a very funny effect as the inflated balloon slowly appears on the slate.

Magic Hair Restorer

You draw a picture of a bald headed man, then sprinkle it with a little of the magic liquid, (after laying it down) then go over with a moist cotton wad. It now reappears, but with the bald head covered with a large crop of hair. If you prefer, use red chemical chalk to cause the hair to come out as red hair.

Other Ideas

Representation of the famous Indian Rope Trick. Boy climbing rope. Slate washed, boy vanishes.

Houdini escapes from milk can. First picture shows Houdini inside milk can. Washed off and he is standing alongside of milk can.

A picture of a rose bush with buds is shown. Washed off (the buds only) and as the picture slowly develops, the bush is seen to be blooming with beautiful roses. The bush proper should be of the chemical chalk drawing. Make the roses red and the leaves green for a beautiful colored chalk drawing.

Some other Slate-Writing Techniques

History of magic has it recorded that there were always several ways to do the same magical effect. Research uncovered several other ways to perform visible slate writing.

By way of explanation, most any chemical reaction in which two solutions, both colorless, when mixed will form a visible color, may be used in this effect.

Using a silver nitrate pencil to secretly write your message on a plate, it may be developed by several methods. However, it is first necessary to make it invisible. Merely breathing on the writing will make it disappear.

To develop this writing, wash it with a strong (saturated) solution of ordinary table salt. This causes a white writing of silver chloride to appear.

Another method is to first sketch into a slate (very slightly) the message with sulfuric acid. If properly done it will be invisible at a short distance. Washing this with a solution of lead sub-acetate will cause the writing to appear.

A still third method originally performed in the latter portion of the 19th century and early of the 20th century is the following method. It defies detection after the writing appears on the slate and is perhaps one of the best.

First wash the slate with water (real water) in front of the spectator. Prior to the performance prepare a newspaper with the message on it (mirror image—will understand as following is explained). Use ordinary school chalk (white or colored). The way to do this is to first draw the message on a slate and holding it up to a mirror, copy the reflected picture onto the newspaper.

Wrap the washed wet slate with this paper and then cover with a silk.

Have spectator sit on the slate as you continue on with the rest of the routine. The pressure from sitting on it causes a transfer of the message to the slate and with subsequent revelation, a terrific effect is created.

Chapter **XI**

Chemical Cocktails

The Magic Soda Fountain

(Earl C. Leaming)

Here is a method for producing from water—lemonade, orangeade, green river, "milk," grapeade and "beer."

Prepare either one of the following two solutions and fill a pitcher with it.

Simple syrup	20 ounces
Water	68 ounces
Citric acid	1 ounce

or

Water	120 ounces
White Karo Corn Syrup	1 can
Saccharin tablets	enough to sweeten
Tartaric acid	25 Grams
Citric acid	25 Grams

Prepare seven plasses and one beer stein as follows:

No. 1. Unprepared (water)
No. 2. Orange dye and 5 drops orange essence
No. 3. Lemon yellow dye
No. 4. Green dye and 5 drops lime essence
No. 5. Purple dye and 5 drops grape essence
No. 6. A few drops of sulfuric acid
No. 7. A few drops of a lead acetate solution
No. 8. A small amount of "Beer" powder

In the beer stein—if the "Beer" powder is not available—the following may be substituted: one-half teaspoonful sodium bicarbonate, a few drops of a carmel solution and a little saponin powder.

Proceed as follows: Produce the pitcher of solution, calling it water. Have your assistant bring forth the tray of prepared glasses. The bottoms of the glasses are concealed by the edge of the tray, but a few unprepared glasses amongst the prepared ones on the tray are useful since they can be exhibited and then switched for the prepared glasses.

The first glass is filled from the pitcher and drunk to prove the use of "water." The second to fifth glass give beverages as explained in the first portion of this routine. These can also be supplemented with other prepared duplicates, depending on how much time you have.

No. 6 is again "water" which is placed back on the tray. Using any suitable means to cause "milk" to be mentioned, pick up this glass and pour into No. 7. DO NOT DRINK THIS MILK!!!

Use the solution in the pitcher to produce "beer" in glass No. 8. It is well to bring on the stein upon a separate tray, since there is a chance that foaming over might cause stains upon the table or floor that may prove undesirable.

Simple syrup is made by dissolving as much granulated sugar as possible in water to make a thick clear syrup. It may be purchased ready made from most drug stores.

Yellow color may be TARTAZINE YELLOW or "lemon shade" certified dyes.

Orange color is prepared by mixing yellow and red certified dyes. CLARENDON ORANGE (American Aniline Products, Chicago) may also be used.

Purple color is a mixture of blue and red certified dyes.

Green color is "mint shade," "lime shade" or brilliant green certified dye.

Flavor essences are available from the Liquid Carbonic Company or from any of the many "magic" dealers throughout the country.

Sensational New "Beer Powder"

It is now possible in any trick in which you use water, to produce a foaming glass of beer as a smash climax.

At the finish of some regular water trick, you can have some "stooge" (possibly one of the musicians in the orchestra, a waiter, or an usher) call out, "Now let's see you change that into a glass of BEER!" There is always a roar of laughter at this because the audience thinks the performer has been "stumped." Pause a moment, look confused, letting the "point" sink in. Pick up the vessel containing the water (not too cold) and pour the water into the beer glass in which you have the "Beer Powder." Seemingly you have produced a foaming glass of beer.

Use about a level spoonful to an ordinary eight ounce tumbler if the water is at room temperature. If the water is ice cold, use fifty percent more powder.

Important: Pour the water in a thin stream, from a distance of about ten to twelve inches above the glass. Wiggle it or make the stream run back and forth over the powder so that the water will serve to stir the powder continually while pouring. As you fill the glass and reach the half-way mark, stop for a second and then continue to pour the water. This keeps knocking the foam down into the glass and serves to make a richer and "finer-grained" foam by washing back into the liquid any coloration that might get carried to the top. A little practice will show you how to make a foam that is "creamier" and "whiter" in appearance.

Keep the powder dry at all times! Keep the cover on tight! It is highly important that no moisture be allowed to seep into the container.

Mr. Burling "Volta" Hull who developed this powder can supply it, or you can get it from any "magic" dealer.

The Magic Soda Fountain

With permission by Burling "Volta" Hull, the patter and

presentation are reprinted here for this effect. The new directions were copyrighted in 1937. All rights for manufacturing are reserved by Burling "Volta" Hull.

One of the secrets of this unique and novel magical mystery is based upon certain recent developments in the science of flavor compounding which has lately produced a new line of super-concentrated fruit and soda fountain flavors unlike anything known previously. Flavors so amazingly concentrated that it almost baffles comprehension. Imagine, if you can, a flavor so concentrated and condensed that one glass of it will flavor five thousand and forty glasses of soda water with the same identical strength and richness of flavor which you are used to getting in your regular soda fountain drinks. To make it easier to understand, one single drop will flavor two and one-half glass tumblers full of soda water. Of course, such tremendously concentrated flavors would not be instantly soluble in plain water so they have to be prepared by means of a special type of recently developed solvent. Thinned with this solvent, only a few drops or so are needed to flavor an entire quart. Thus for each glass as used in this effect, only three to five drops are needed.

An advanced type concentrated sweetener, developed in the chemist's laboratory (over five-hundred times as sweet as sugar), requires only four or five drops in an entire pitcher of water to provide the usual sweetness you are accustomed to in the regular soda fountain drinks is employed.

The use of a modern scientific flavor solvent developed specially for these highly concentrated flavors which renders them instantly soluble in either plain or carbonated water is used.

A fourth secret is the use of the latest development in a line of super-concentrated food colors (U. S. Government Certified Pure Food Colors) so astonishingly concentrated that in some cases only one-fourth of a single drop is sufficient to impart the proper coloring to a glass as found in the regular soda fountain drinks.

By use of a special ingredient recently developed, a "creamy" appearance is instantly imparted to the drink so that it resembles that milky or creamy appearance familiarly associated with either plain cream or ice cream sodas.

And finally the sixth necessity was developed to produce a delightful effervescense and creamy foam without the use of the regular high-pressure soda carbonating equipment. It is obtained by the mere act of pouring the water into the glass.

Patter and Presentation: "Here sonny, you're looking a little pale after all your hard work helping me up here. Here is a nice strawberry soda. That ought to give you a little color in your cheeks.

"And now this little girl, we'll give you a nice golden vanilla to match your golden hair.

"Now this boy over here, you're too anxious, I guess we'll just hand you a lemon!

"Here sonny, you like apples? Yes—well that's good. Here is a pineapple soda.

"Little girl, here is something that is good for what is 'ailing' you! A nice sparkling ginger ale . . . that will put ginger in you.

"Come, son, what are you shaking your head about? I guess it is just a case of sour grapes with you. All right, I will give you a grape soda. Now doesn't that make you feel better? Aha, I thought so. Yes, he says that tastes just like grape.

"Now I will take this big husky chap over there. He will want something strong. How about a glass of beer . . . I mean root beer! That will put hair on your chest.

"This little sassy chap who was talking all the time during my last trick, I guess we will have to give him something in a sassy drink. Here is a 'sassyparilla.'

"That other fellow over there, you needn't feel so blue, sonny. I guess we will have to give you a nice 'cheery' drink—here is a nice cherry drink.

"Right over there, that boy with the orange colored hair, I will pour you a soda to match your hair, an orange sode. Real Sunkist orange (holding up the glass and looking at it) of course.

"And that chap over there, whose father is in the brick-laying business. I'll give you a little lime to go with your bricks. That's right, just try it on your tonsils first.

"Now my friends, don't all call at once. Don't razz me for more drinks. As this is my last drink, I'll just hand you a raspberry."

A good stunt to finish with, is to pour out a glass of "beer" for the performer. The performer can figure out his own gags to go with each drink for effective presentation and thus make the routine individual with himself.

Method of Working: The performer places in each glass a few drops of these highly concentrated flavors in the bottom. He then spreads the flavors evenly over the entire bottom of the glass by shaking the glass gently. So prepared, these few drops of liquid are not noticeable in the glasses at the customary distance between the audience and performer. To still further assist the illusion, the performer can use a vari-colored silk or cloth laid over the tray on which he places the glasses. The mottled and many colored design of the cloth makes the minute quantity of flavors indistinguishable. In fact, the performer himself, who known what to look for, may have difficulty in recognizing which glass is which. It's well to lay them out in a definite pattern in such a way as he intends to use them.

The various flavoring extracts (essences) may be obtained from either Burling "Volta" Hull direct or from your favorite magic dealer.

Pouring the Drinks: The "water" that is used is really carbonated soda water, which can be bought for a few cents a bottle in any drug store. For best results either have this cold at the beginning of the performance or place a large piece of ice in the glass pitcher (which may be borrowed) and then pour the soda slowly against the side of the pitcher. Tilting the pitcher slightly, pour soda water against the side so that it runs in slowly without being stirred up too much. Do not shake or jar the pitcher any more than you can help, as it may make the water lose its effervescence and become flat. The ice will keep the water cold during your performance until you are ready to use it. If the water becomes warm, it loses a large part of its fizz. Needless to say, the drinks also taste much better when they are cold. The pitcher now will appear to contain only plain ice water.

In picking up the glasses, disregard the preparation of same, because the few drops of concentrate placed therein will not be seen.

Into the glass pitcher or bottle you are going to use, put in the proper number of drops of concentrated sweetening on the following ration: two drops for each tumblerful of water that the pitcher will hold. Do this before pouring the carbonated water in, so that when you do pour in the water, the sweetening agent will be automatically mixed in all through the carbonated water avoiding the necessity for stirring.

To pour a lot of drinks and make more glasses of soda is accomplished by the method of pouring. Use small size glasses. To pour a foaming or creamy soda, start pouring about an inch of the liquid into the glass, raise the pitcher with a flourish, till it is about six to eight inches above the glass so that the liquid falls into the glass with more force. This causes the soda water to foam up more, producing a big head of delicious foam at the top of the glass. Incidentally, it fills up the glass, thus enabling the filling of more glasses with soda; thus making the liquid go further. Four brimming tumblerfuls of water (equal to about one quart) poured in this way will give you about twelve glasses of creamy soda, as you only use about one-third of a glass of actual liquid, the rest of the glass being filled up with foam—just like at a regular soda fountain.

A word or two about the "creamy" ingredient. When desired to simulate a "creamy" effect in sodas, place two drops of the special creamy ingredient available in each glass at the same time you put the concentrated flavors in. It makes for a better contrast if you have both clear and creamy drinks coming from the same pitcher or bottle.

Also, it makes a nice touch for a children's entertainment to have a small dish of ice cream and to add a spoonful to each soda.

Chapter XII

Chemical Stunts

A Chemical Smoke Screen

Heating ammonium chloride will give off thick clouds of smoke which will continue until the substance is entirely volatilized.

Magic Ice

Prepare a saturated solution of sodium bisulfate in water and to this add an equal volume of a similar saturated solution of sodium silicate.

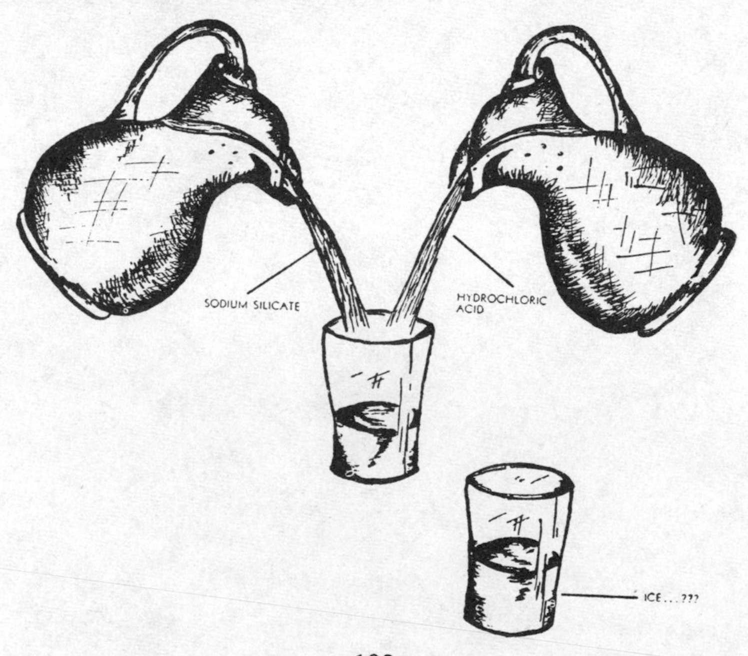

The mixture will gradually "freeze" into a transparent solid, becoming hard at first at the surface and gradually solidifying through the entire mass. In a minute or two the container can be inverted as there will be no liquid to spill.

How to Make Bones Elastic

Obtain a small chicken bone such as the wish bone. Clean it thoroughly and cover it with strong vinegar for twenty-four hours. At the end of this time remove the bone from the vinegar and it will be so elastic that you will be able to tie knots in it without breaking it.

How to Stretch an Egg

Obtain an egg and put it in a glass full of vinegar and allow it to stand for about twenty-four hours. At the end of this time feel the egg with your fingers and if it is soft and elastic, remove it from the vinegar and replace it with fresh vinegar for another twenty-four hours.

When the egg has become elastic enough, it may be put in a bottle through which it originally would not pass without breaking. An elastic egg will be quite a novelty to show your friends, but be careful not to break it.

The Magic Handkerchief

Moisten a small handkerchief with cobalt chloride dissolved in water. Dry the handkerchief on a heater and you will find that it is blue when thoroughly dry and warm.

Here is where you have some fun.

Show the handkerchief to someone while it is blue, and then crush it between your hands and blow through it for a few minutes. It will become colorless. This can be repeated as often as you like.

Green Violets

Heat in a suitable container equal portions of ammonium chloride and calcium oxide over a flame. Hold a blue violet at the opening of the container as you heat it. The gas which comes off will turn the flower green.

If you try other kinds of flowers you may obtain some weird and unusual results.

An Obedient Egg

Obtain two one-quart fruit jars. Fill one of them three-fourths full of water and the other three-fourths full of a concentrated salt solution.

Now place an egg in the jar of water and it will sink. Place it in the salt solution and it will float. By remembering which jar contains the salt solution and which the water, you can make the egg obey you and either float or sink as you command.

Water That Will Not Spill

Obtain a piece of cardboard or stiff paper large enough to cover a tumbler. Fill a tumbler brimful of water and slide the piece of cardboard over it. Now hold the card in position and turn the tumbler upside down. You will find that the cardboard no longer needs to be held as it sticks to the glass due to atmospheric pressure, and this holds the water in the glass.

The Mysterious Mothball

Fill a cylinderical container with vinegar about three-fourths full. Add some sodium carbonate. Now drop in a mothball and the bubbles which form on it will carry it to the surface. It will then turn around and sink to the bottom again. This action will be repeated again and again.

The mothball must be naphthalene.

Chapter XIII

Professional Magic

Patriotic Liquids and Silks

(Walter Essman)

Effect: The performer displays an empty cocktail shaker . . . he also shows three silks; red, white and blue. The silks are put into the shaker and the cover put on. (The cover has a pouring spout.) A tube, slightly larger and similar to the

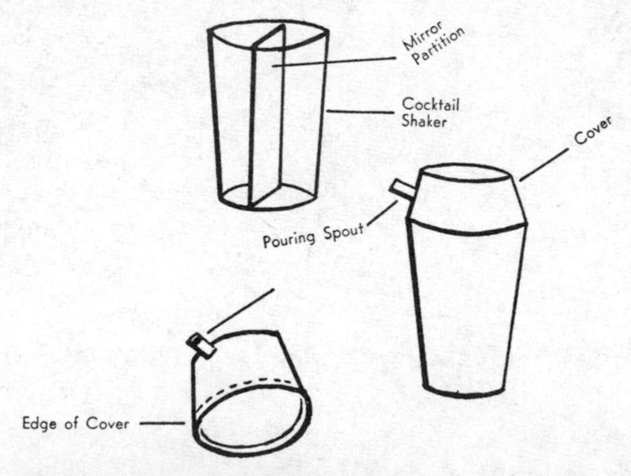

shaker is also shown. The shaker is put into the tube with the cover showing at the top of the tube. Three empty cocktail glasses are set in front of the covered shaker. Picking up the shaker and tube the performer pours a red liquid into one glass. Into another a white liquid and into the third a blue liquid. Removing the cover, a large American flag is produced. The shaker and tube are shown empty at the conclusion.

Method: The cocktail shaker is the glass type. The shaker is divided through the center with a mirror partition. To make the mirror divider, lay the shaker on the side on a piece of cardboard. Trace around the shaker at its widest point and cut out this pattern of cardboard. Trim the sides of the cardboard until it fits snugly in the shaker. When the pattern is correct, trace it onto a piece of highly polished tin. Cut out the tin form carefully with a pair of tin snips. Insert this in the shaker and glue into position.

The cover is next prepared. If your cover does not have a pouring spout, you may make a hole near the top which will serve the purpose. Cut out a piece of aluminum or any non-rusting metal so that it will fit into the bottom of the cover . . . about one-quarter inch from the bottom edge. This is soldered to the cover.

To each of the glasses you place two drops of vegetable dye of the required color. The white is any of the imitation milk liquids sold by the "magic" dealers. The tube is made from heavy cardboard and bent around the outside of the shaker. Cut it so it will fit perfectly. Decorate the tube with alternate stripes of the colors used in the effect.

Conceal the flag in the rear portion of the shaker. Fill the cover with water through the pouring spout. You are now set to "mystify your audience."

With the empty front part of the shaker showing, put the three silks into this part . . . put on the lid. Lower the shaker into the tube. Pour water into each glass, the dye in each glass will turn it the desired color. Before removing the tube, turn the shaker so the flag is facing the audience. Remove the tube and cover . . . pull the flag from the shaker with a flourish.

Magic Dyeing

(Earl C. Leaming)

Apparatus: A glass cylinder. This is a so-called "Hydrometer Jar," ten inches by one and one-half inches. If desired, the lip or pour-out may be cut off, so as to have a smooth top and a height of about ten inches, a phantom tube, which is a regulation piece of magical equipment. This is a lacquered tin tube, three and one-half inches in diameter, with an inner tube in the shape of a truncated cone tapering to two inches in diameter at the top and lacquered dull black inside. A water pitcher, three tumblers, a red, a blue and a white silk (separate silks) and a silk American flag, or a rainbow colored silk.

Routine No. 1: Prepare the water pitcher with plain water and add one teaspoonful of a saturated solution of sodium carbonate to each quart of water. Prepare the glasses by placing in the first, one teaspoonful of a solution of ortho-cresol-phthalein; in the second, one teaspoonful of a strong solution of citric acid; and in the third, one teaspoonful of a solution of thymolphthalein.

Call attention to the pitcher of clear water and the three "empty" glasses. Announce a new method of dyeing. Fill the glasses with the water from the pitcher, producing red, white and blue colors.

Now show the cylinder and phantom tube. Cover the cylinder with the latter, pour equal portions of the liquids into the covered cylinder and proceed to produce the colored silks from the phantom tube. Then remove the phantom tube and show the cylinder which now has clear water, the colors having "vanished."

Routine No. 2: In addition to the apparatus just used, prepare a celluloid cylinder having red and blue bands painted on it.

This cylinder fits over the glass cylinder projecting over the top about one-fourth inch. In use, the celluloid fake is concealed in the phantom tube and placed over the glass cylinder at the beginning of the presentation. After pouring the colored liquids into the glass cylinder, the phantom tube is briefly raised, seemingly showing the various colored liquids in layers. When the phantom tube is again removed after producing the colored silks (or rainbow), the celluloid cylinder is removed along with it in the same manner as the tubes in the "Passe Passe Bottles."

From here the routine is similar to that in the first routine.

Several variations are possible. One might end by producing a glass of beer using any of the several methods explained elsewhere in this book.

Califaction of Handkerchiefs

Load a thirty-six inch silk (or as many silks as you can use) in the gimmick of your dove Pan. Set aside in the usual manner. Show the pan empty and place a small piece of flash paper to one side. Next place several twelve inch silks in the pan itself.

Your patter might present the effect to the audience by way of explanation that heat is needed to blend the colors together. Ignite the flash paper and quickly put on the cover.

Remove the cover and draw forth the thirty-six inch silks (more effective to use a six foot square silk, rainbow colored). Show to the audience that you have actually califacted the single silks together. (This is one of the many several effects one can expect to find in the Ireland Year Books, and grateful thanks for permission to reprint them is expressed.)

Mystic Papers

This is the standard effect where a strip of paper is folded in half, cut and, when opened, it is found to be restored in one piece.

Prepare a strip of paper (newspaper is best) about two inches wide and twenty-four inches long. Coat one entire side with rubber cement and allow to dry. Dust with talcum powder over the coated side and shake off any excess.

Fold the paper in half, and with a pair of sharp scissors

clip off the folded end. Upon opening up by taking one edge opposite the end cut with the scissors and allowing the other

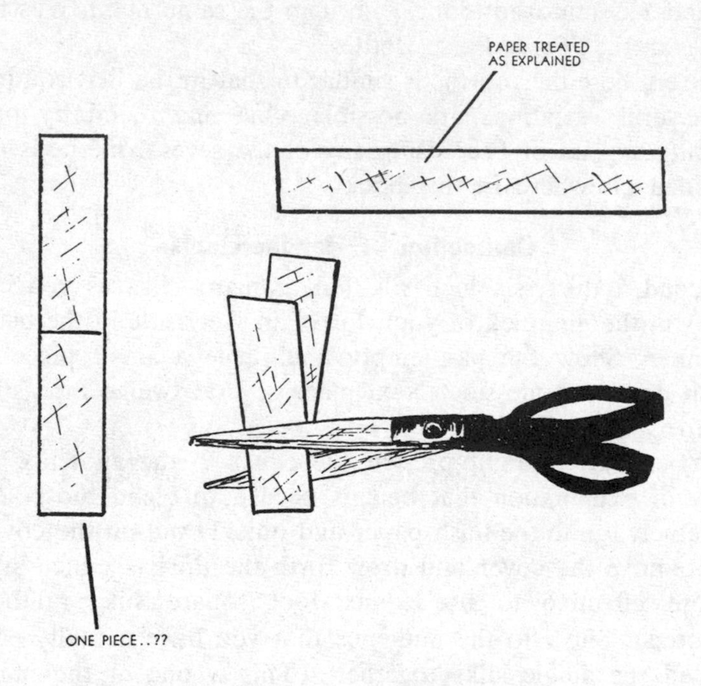

edge to drop downwards, the paper will appear to be in one piece. This is because of the two edges of rubber cement, when cutting the paper, will make an invisible joint.

These papers may be prepared in advance.

The Bewitched Paper

(Dr. Howard B. Kayton)

Someone from the audience is asked to come upon the stage to help in an illustration of the motto, "If two do the same, it is not the same."

Displaying two sheets of paper and two scissors, the volunteer is given the choice of one of the scissors and of one of the papers. He is requested to proceed exactly as the magician.

Cut a strip about one and one-half inches wide from the long side of the paper. Keep the narrow strip and discard the rest. Fold the strip in half and tear it along the fold, making two strips. Display them both, holding one in each hand demonstrating to the audience there are now two pieces of paper. Next place the papers over on top of one another and cut the edges so that both pieces are exactly alike in size and form. Make a movement as if to take one piece of paper in each hand and instead of being separated, the two edges are found to be joined! They again form the original strip, whereas the parts of the strip held by the spectator remains separated.

Take the pieces from the volunteer, go through the same movements as before and again the papers are in one long strip like the original.

The method is simple. Both papers are prepared. The first one is prepared with rubber sement and talc one-half inch wide at the center. The other is prepared by coating the top and bottom edge of the paper in the same way.

The performer has only to watch that the volunteer folds the strip with the preparation on the inside. By this means the volunteer who has executed the same actions as the performer will never be able to restore the paper strip as the performer does.

Stage Size Out to Lunch Trick

(Ireland Year Book)

Magician exhibits a large picture (about 8 by 11 inches) which shows a Hindu boy climbing a rope, which rope emerges straight up out of a basket. (Same idea as the sketch on the pocket trick "Out to Lunch.") A spectator writes his name or some other mark of identification, in or near the basket. The picture is turned over and spectator marks reverse side. On turning picture book to original side an "Out to Lunch" sign is seen where the climbing boy was before.

Method: A clip board large enough to take the size picture being used is required, in this case a little larger than 8 by 11 inches. Face of the clip board must be black and coated with rubber cement. (In actual use, coating entire face of clip board with rubber cement is not necessary. An inch around periphery to correspond to size of fake will be found satisfactory.

Picture itself should be yellow, on a black piece of cardboard. Thickness of cardboard is optional, but business card stock is fine. Fake consists of just upper half of picture (yellow drawing on black background). Fake is held to main picture by daubs of magician's wax. Face of fake is coated with rubber cement about one-half inch around periphery. In presenting the trick, picture with fake attached is clipped to board, the better to display it, and allow spectator to sign his name in basket. Picture is turned over for spectator to write on back. Fake will adhere to clip board, but will not be seen because clip board is black and fake is black. Therefore, when picture is turned back, boy has vanished and "Out to Lunch" sign is seen in its place.

White chalk is given to spectator for the signing, and can thus be wiped off so as to use the trick again.

The Chemical Garden

(Earl C. Leaming)

If science continues to develop as rapidly in the future as in the past, man will not depend on the methodical but snail-like Nature to produce life. Chemistry will supplant her with the aid of diabolical mixtures which will be able to produce living things. Why should we plant seeds, patiently wait for germination to takeplace, care for the delicate seedlings, only to have them killed by the frost, burned by the sun, or mowed down by hail, when the chemist can produce and deliver the finished product in a twinkling?

These and similar baseless opinions are often heard among all classes of people. But, it is an absolute fact that plants can be produced artificially, although not as the fantastic hopes of the optimistic have pictured. That which science has produced resembles the plant in many respects. Stems, leaves and even fruitlike forms may be seen, but the vital spark of life is absent. They are as cold and dead as stones. The divine touch of the giving of life has not yet been entrusted to man. Science can, however, seemingly produce life. A life so realistic that the uninitiated are amazed with wonder.

Such an effect can be produced with the chemist's flower garden. A chemical seed is dropped into a clear transparent liquid and after a moment it begins to germinate. Watch! Already a long shoot has developed—it grows momentarily longer—it stops, seems to hesitate, when suddenly a protuberance is formed. The fruit. Another shoot has meanwhile reached the surface of the liquid, begins to spread and a leaf or two has been developed.

The method: The liquid used is ordinary water glass, about one part of this liquid in one part of water. If the solution is muddy or turbid it should be filtered. If a high grade of water glass solution is purchased, it should be fairly clear. For seeds, small crystals are taken of the following substances: Cobalt nitrate, a red crystal which slowly changes its color to a chlorophyll green as it grows; nickel nitrate, which produces fine tendrils very rapidly; manganese sulfate, which produces other shoots; and ferric chloride, which develops thick gnarled twigs and thin laminated shoots from which leaves soon begin to grow. Aluminum sulfate forms thin delivate shoots.

Cobalt chloride, copper nitrate, potassium ferrocyanide and calcium chloride all give different effects. Nitrates and sulfates work best. These plants are all one celled as are the natural plants found in ocean life. Small containers with only a few crystals are better than large masses. The phenomenon is a problem in osmosis.

Magic Garden Seeds

Mix six grams copper sulfate, one gram iron sulfate, six grams manganese sulfate, one gram cobalt chloride and four grams plaster paris. Mix with sufficient water to make a thick paste. Mold this paste into the shape of beans and allow to dry.

Place one in a solution of sodium silicate and water as outlined in the foregoing effect. Set aside and watch the results. A similar chemical garden will shortly begin to grow.

Dry or Wet

Hands placed in a bowl or water are completely dry upon removal. When a spectator places his hands in the same bowl of water, his hands are dripping wet.

To accomplish this, purchase some powdered lycopodium. Secretly dust the hands with this powder and shake off the excess. Lycopodium has the property of shedding or repelling water.

Anti-Climax

A number predicted (by your favorite method) is made to appear on a blank card when ashes are rubbed on it. Secretly write with soap on a white card the numbers which will be invisible until the ashes are rubbed on.

Flaming Beauty

From an idea suggested by Lloyd Jones, this very effective magical gag or M.C. bit was evolved.

Prepare a flaming mixture (see phosphorous and carbon disulfide "Fired Liquid" mixture in chapter on Chemical Magic with Fire).

Secretly wet a piece of cotton with this liquid and call a girl over from the audience. Tell her, "Honey, breathe on this for me" (cotton a glass dish). It will suddenly catch on fire and then follow up by remarking, "She's sure hot stuff!"

Reading Sealed Envelopes

Secretly have a piece of cotton saturated with ether conveniently located for easy access. Obtain the envelope with the message sealed inside.

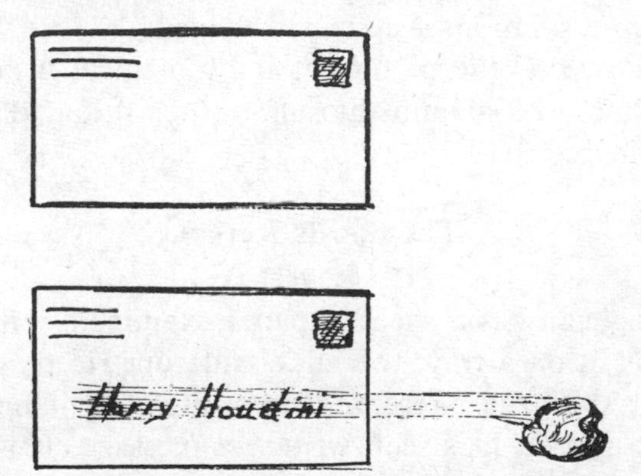

"Steal" the cotton and rub across the envelope. You will be able to read the message in a few seconds and the ether will evaporate almost immediately leaving no tell-tale marks.

Another Liquid — ?? — Solid

Equal portions of saturated solutions of calcium chloride and potassium carbonate in water, when mixed will form a solid mass.

Writing With Fire

Prepare a solution of potassium nitrate in water (saturated). Dissolve a small amount of gum arabic in it. With a regular pen write your intended message using this solution as ink. Set aside to dry. Touch a lit match to a spot on the writing and if the lights are extinguished, the fire will be seen to conform to the invisible writing (unknown to the audience). Thus is revealed the hidden message.

Spirit Photograph

Your own imagination will undoubtedly devise a potential routine for the photographic effect to be described. Its use in magic is unlimited, so guard its secret with care.

Develop your selected picture on "printing-out" paper. Your photographic dealer can furnish you with information.

Thoroughly wash the print in distilled water and then bleach it in a solution of mercuric chloride.

To "redevelop" the picture as if by magical means, place it in a solution of sodium thiosulfate (hypo) and the original picture will return.

The Spirits Reveal

(E. L. Palder)

The magician has a sheet of paper examined, initialed and then places it on a tray, the initial side up. He than goes on with the rest of the effect and upon finishing it he removes the paper and on it is seen written a message. It is the predicted answer as determined by the particular effect the magician might be doing.

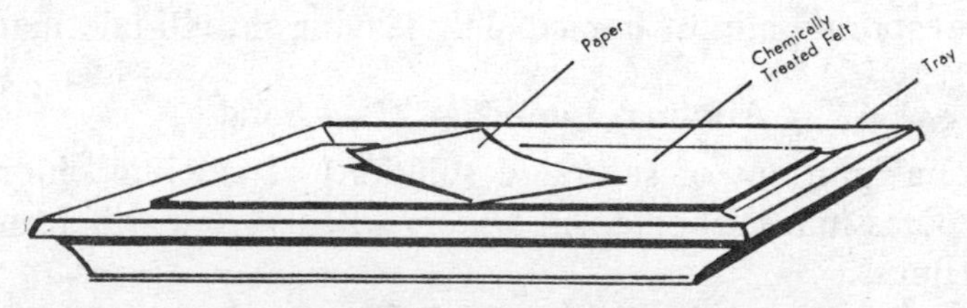

The fact that the paper, after being initialed is placed on the tray held by the spectator, and not touched by anyone, places an important psychological thought in the minds of the spectators. No switching of papers. The use of a spectator holding the tray precludes any possibility of switching throughout the entire effect.

To perform this effect first obtain a tray with a slightly depressed bottom. Cut out from black felt and glue a piece the exact shape of the bottom in place with rubber cement. Moisten the center of the felt with a solution of tannic acid in water. The papers should be prepared in advance by writing the intended "predicted" message with a solution of potassium ferrocyanide and set aside to dry. The tray should be moistened just prior to the performance.

The procedure is obvious. Contact of the paper with the tannic acid solution causes the message to appear.

Flash Pot

At your local metal shop, have made a metal container with a bottom six inches in diameter and the sides conical in shape, five inches high. The top should be open (no top) and have a diameter of nine inches.

Obtain a two prong male plug (electrical) and fasten the plug, prongs up, to a block of wood one-half inch thick. This block of wood should be about one inch square. Drill a hole through the center of the wood and through the bottom of the metal container. Fasten the wood block with the plug to the pan and then in the conventional manner wire the plug with insulated wire. Pass this wire through the pan and connect it to a second male plug. This is the plug you will insert into an electrical outlet later on.

Fasten a piece of thin, single-strand, and not insulated wire across the prongs of the plug in the pan.

Obtain some magician's flash powder and place a small quantity on the plug in the pan covering the thin strand of wire.

Step back; turn on the current and "poof" you have a smoke screen (without fire) for use in magic.

Dark Eyes

(Teral Garrett)

Into each of four examined, white drug-type envelopes, each of four spectators place an ace. They seal the envelopes and then thoroughly mix the four envelopes. The lights are extinguished and any one of the four envelopes is handed to the performer and he, in total darkness announces which ace is inside. When the lights go on once more, the envelope is opened and therein is the ace named.

Before presenting the effect, place about six or seven grains of salt in one envelope, the same amount of sugar (saccharin is better) in another, in the third about half of a pulverized

aspirin tablet, and the fourth one is left unprepared. The envelopes thus prepared may be freely handled and the spectators permitted to insert their respective cards and seal them. All that must be remembered is which spectator gets which envelope for the Ace of Clubs goes into the bitter (aspirin). the Ace of Diamonds in the sweet (sugar), the Ace of Hearts in the salty (salt) and the Ace of Spades in the blank (empty) envelope.

When one of the envelopes is handed to you behind your back, with the point of a knife blade, make a small incision in the corner of the flap, cause the envelope to bulge open a little by exerting pressure on the corners and permit the salt, sugar or aspirin to pour into your hand. Taste it, put the knife away, and announce the name of the ace and call for the lights.

As the lights are turned on, hold the envelope between the thumb and forefinger so that the incision is covered. Bring out the knife and insert the blade in the hole you made and cut it open entirely. Remove the ace to verify and then everything can be examined.

(Editors' Note: This method allows the spectator to mark his envelope with an identifying mark so when the lights are turned on, there will be no cause for anyone to suspect switching of envelopes, which of course, there wasn't.)

Revelation

The magician exhibits a piece of cardboard and shows both sides to be free from writing or any marks of any kind. Holding the cardboard, he has a spectator initial it. He then places it to one side in full view.

After completing the rest of the effect such as a message or number prediction, etc., the cardboard is turned over and the message mysteriously has appeared on the other side.

This is an old effect. Its origination has been lost in the antiquity of magic and here it is presented with two improvements.

Method: Two pieces of cardboard are held as one. The inside surfaces are treated with roughing fluid. The message to be "revealed" is already written on the inside of one piece of cardboard. It is the other side of this cardboard that the magician has initialed.

After being initialed, the cardboard is placed on a metal tray to which a small piece of magician's wax has been placed. The cardboard thus will adhere to the tray if pressed down onto the wax.

To reveal the message, pick up the tray, break the seal between the two pieces of cardboard and drop the cardboard off into the hands of the spectator by dropping the tray forward and downward. This action brings the bottom side of the tray towards the spectators with the topside having the other piece of cardboard fastened to it facing the magician.

With a little shownmanship a mystery well deserving of magic may be accomplished.

Prediction

(E. L. Palder)

Make a tray using three pieces of plywood one-fourth inch thick. Cut them in thc following manner: two pieces approximately eight inches by eleven inches and one piece the same size with a section a little larger than a standard size envelope cut from one side. Glue this piece between the other two, making a tray with a hidden slot in one end. You can secretly secure an envelope here unknown to the audience.

Fasten some magician's wax to the top of the tray. Prepare an envelope with your prediction (headline of the newspaper of the day). Have notarized by a friend of yours who will back date it and then you may secrete this envelope in the slot in the tray.

This is all done just prior to the show. However, some days before, even weeks, you have taken an envelope, prepared a prediction (which of course is not true, nor do you allow any of the committee present with you to see.) This is sealed in an envelope and notarized by your friend again. This envelope is retained by the committee until the night of the performance when it is brought forth and handed to the magician on the tray. A member of the committee is brought up on the stage who relates to the rest of the audience what has transpired prior. With this, the envelope is dropped off the tray into the hands of the spectator (envelope in slot, the other envelope is stuck to the tray by virtue of the magician's wax thereon).

The envelope is opened, the prediction read and a miracle is created.

Misdirection

From an idea for a top notch routine using chemicals from Bob Carver of Macon, Georgia, comes the following:

Set up in the following glassware in the order listed with preparations as indicated. The list starts from left to right.

(1) Empty shot glass

(2) Clear glass pitcher filled with water

(3) Empty glass large enough to hold the shot glass

(4) A second glass as in No. 3

(5) Clear glass pitcher filled with tetrachloroethylene

(6) A second empty shot glass

Effect: Patter about not letting your left hand know what your right hand is doing. Show glasses to be unprepared (without saying so). Pour contents of pitcher into glasses at the same time (one hand mirroring the other's actions). A light set up in back (can be eliminated) lets the audience see "through" the glasses. Drop the shot glasses, one into each glass. One vanishes completely, the other is still in plain sight (in fact the water will seemingly magnify the size of the glass).

Due to certain physical properties of a highly technical nature, tetrachloroethylene has the property of making a transparent object invisible when place in it.

Tetrachloroethylene is available from dry cleaning supply houses. Do not get "Tri" chloroethylene which will not give the same illusion and does not appear like water.

Bottle of Smoke

(Ireland Year Book)

An empty bottle, preferably a quart round whiskey bottle is filled with smoke with the aid of a soda straw. Use several mouthfuls of smoke. The smoke will not come out. No one in the room can make it come out easily in less than five or ten minutes. Yet the performer challenges that he will empty the bottle in two seconds and without touching the bottle.

Have at least a teaspoonful of the whiskey or other high proof spirits in the bottom of the bottle. A whiskey bottle that has just been emptied always has that much in it. Fill the bottle with smoke, then handle the bottle with a rolling motion, so as to wash the bit of whiskey around on the bottom and sides and get the fumes throughout the smoke. Now, after being thus prepared, sit the bottle on the table and make the above challenge. To prove it, drop a lighted match into the bottle. The fumes ignite and the smoke burns (or appears to burn) instantly and with a beautiful blue flame. Use plenty of smoke.

Production of Three Large Balloons From a Paper Cone

Performer takes up a sheet of paper from his table, shows both sides and then forms a cone. Holding the cone with his left hand, he dips his right into it and a few moments later a large balloon is produced. Then a second balloon and finally a third.

I

The paper is about thirty inches square. A thread is attached to one edge of the sheet, at the center. The thread should be eighteen inches long between the paper and the load. The load, of course, is three prepared balloons. When the paper is picked up, the edge with the thead attached is at the bottom, level with the table top. Load is on servante. Sheet of paper is turned over, top falling forward. This brings the load behind the paper, right in the center. The cone is formed, and the load, now inside the cone, is snapped off the thread.

Each balloon should be of heavy rubber and is prepared this way. First take three tablespoonfuls of baking soda and moisten down into a thin paste so it can be flowed into the balloon. Fill a small bottle, about one and one-half inches tall and three-eighths inches around, with commercial sulfuric acid. Cork with rubber stopper. Place the bottle inside the balloon and tie the neck of the balloon with thread. To produce balloon put hand in cone, take cork out of the bottle, then as the balloon fills with gas, gently work it upwards so that it will not bind in the bottom of the cone. Take the balloon out. Then do the same thing twice more with the other two balloons.

II

Have three balloons tied together and vested. Show paper and steal the balloons. Leave the balloons inside the cone. Produce all three as explained above.

Another Balloon Production From Paper Cone

Follow routine outlined in the foregoing, however, in lieu of the chemicals used, the following may be substituted.

Place a bit of carbide in the balloon and using a vial of the same dimensions as in the foregoing, fill with plain water.

The gas formed in this method will usually fill a balloon of average proportions fairly fast. (Caution: The gas formed in this case is inflammable, so avoid any fire or sparks when using this method.

And More Balloons

Prepare a mixture (keep dry) of five parts, by weight, of sodium bicarbonate; fifteen parts, by weight, of potassium and sodium tartrate. Sift this powder through as fine a sieve as is available. Any lumps will slow up the gas formation.

Next prepare a saturated solution of tartaric acid in distilled water. A saturated solution is one in which all possible powder has been dissolved.

Proceed as in the other routines outlined, placing the dry powder in the balloon and the tartaric acid solution in the glass vial.

Think! Ink!

(Orville Meyer)

Former versions required the use of two half-full glasses of liquid, one was poured in the other and the reaction took place a few seconds later. The use of two glasses has been eliminated by a special pitcher which is easily constructed by anyone

A ten ounce cream pitcher with fluted or ornamental design is the perfect size, though a larger one may be used if desired. A straight-side, two and one-half ounce glass is cemented, with waterproof cement to the inside of the pitcher. The exact location for the cementing of this glass should be determined by the particular type of pitcher used; it should be in a position so as to keep the two liquids separate and not touch either top or bottom. At any rate, it will be seen that this arrangement creates a double container, keeping each liquid separate in the pitcher until the actual pouring occurs.

Dissolve in warm water the following:

Potassium bisulfate crystals	280 mg.
Soluble starch	456 mg.
Water	120 cc.

Cool and pour into the pitcher. (This solution will be referred to as A.)

Dissolve in warm water the following:

Sodium sulfite anhydrous .... 143 mg.
Potassium iodate (pure) .... 182 mg.
Water .... 60 cc.

Cool and pour into cemented glass inside pitcher. (This solution will be referred to as B.)

If desired the solutions may be prepared at home and carried to the performance in separate containers. However, it is best that neither solution should be mixed more than twenty-four hours before use. When pouring the prepared solutions into the pitcher be sure that there is no inter-mixing.

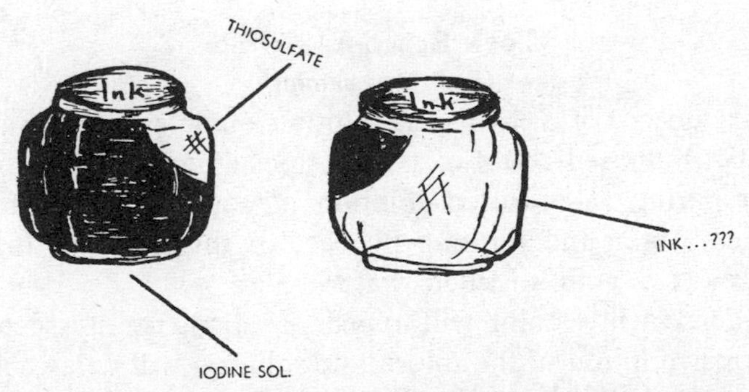

With six ounces of liquid, the reaction will usually take place in about thirty seconds after pouring.

Presentation: Have someone hold an empty glass, and while holding the pitcher (faking cannot be seen at a few feet, and is never noticed anyway) performer explains that everyone is to concentrate on INK—to immagine that the water will soon change to ink. Especially the person holding the glass is to concentrate. Then the "water" is poured into the glass, and the performer continues to say, "Think of ink. We must all concentrate on ink. I feel that someone is not concentrating properly, etc." In a few seconds the water suddenly changes to ink. The performer immediately exclaims, "My friend, you did a very good job!"

Clock Reaction—Yellow

(Earl C. Leaming)

This is another clock or delayed time reaction similar in effect to the blue clock reaction (Think Ink) described.

Take five grams arsenious oxide, twenty-five cc. hydrochloric acid (concentrated) and four hundred cc. water. Boil for five minutes, cool and filter if necessary.

Make a solution of sodium thiosulfate using about three hundred grams to four hundred cc. of distilled water.

Mix equal volumes of the two solutions to produce a bright yellow color change in about twenty seconds.

Clock Reaction—White

(Earl C. Leaming)

Mix ten cc. of a saturated solution of potassium bisulfate with two hundred cc. of a sodium thiosulfate solution (made by preparing a saturated solution of sodium thiosulfate in distilled water and diluting five cc. of this solution to two hundred cc. with distilled water).

A dense white color will appear in about twenty seconds.

Proper dilution of the thiosulfate solution will delay the reaction.

A possible routine for this reaction would be to mix the solutions, hand to a spectator to hold. Light a cigarette and blow smoke towards the glass from a distance. With proper timing, the "water" will become dense in about twenty seconds as if the smoke had filled the water.

The Backslider

(Earl C. Leaming)

A solution shown colored a dark blue becomes crystal clear upon the addition of a few drops of a second solution, but after a time suddenly "flashes" back to blue.

The reaction may be repeated many times.

To eight hundred cc. distilled water add fifty cc., three percent hydrogen peroxide, five cc. ten percent potassium iodide solution, 2 cc. five percent sulfuric acid and twenty five cc. freshly made starch solution.

The resulting solution is dark blue in color due to free iodine and starch.

To cause the solution to "backslide," add a few drops of a fifty percent solution of sodium thiosulfate.

The Money Maker

(Earl C. Leaming)

Make a stamp of wood or metal with a handle and fasten a penny to the bottom. This is used by striking with a small mallet or hammer, in a pretense at stamping out coins.

Take several strips of twenty gauge copper sheet about two inches by six inches, clean them thoroughly with scouring powder or polish, followed by a dip in dilute nitric acid, rinse in water and dry with a soft clean cloth.

To use, rub with a clean cloth pad and mercuric chloride solution. This coats the copper with amalgamated mercury, giving a silverlike appearance.

Pretend to stamp out coins with the fake mold. Substitute a number of pennies previously covered with mercury, and throw them out to the audience.

Patter: "Ladies and gentlemen, I am about to demonstrate for your entertainment the most amazing and marvelous secret discovery of all times. For centuries the ancient alchemists sought for this secret under the name of the Philosopher's Stone, that marvelous substance which possessed the power to turn base metals to gold. Modern science, with its vast arrays of chemicals, electrical power, icroscopes, light rays, and what not, has frantically continued the search without success. It remains for an obscure chemist to discover the formula in the form of a fluid containing a powerful catalyst, which once released penetrates the fabric of of one formula which has the effect of changing copper to metal used and instantly causes the elements to rearrange their structure.

I have the privilege this evening to possess a small quantity silver. I merely take a strip of clean copper, rub in a little liquid, and lo!—we have a strip of pure silver.

I am going to prove to you that this transformation is no trick by stamping out some silver coins for you. There you are."

Magic Flowers

A bouquet of white flowers (or feathers) is made to bloom in colors.

Procure a group of white cloth or paper flowers (or feathers) and make or buy white leaves and stems of paper.

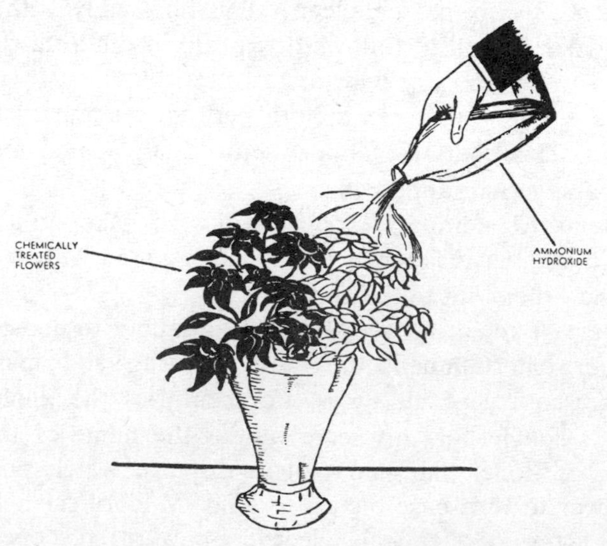

Saturate them by dipping or spraying the blooms in pheno-lphthalein, orthocresolphthalein, thymolphthalein or mixtures of these solutions and use when moist or nearly dry. Coat the leaves and stems lightly with a small amount of finely powdered malachite green dye (dry).

Place the bouquet in a vase on a tray and expose it to the fumes of ammonium hydroxide. When this happens, the change is gradual and quite astonishing and mysterious. Perhaps the best way to accomplish this is to place some ammonium hydroxide in a vase and when ready to perform the effect place the prepared bouquet in the vase. In this manner the effect becomes self-working.

The Smoking Pipes

Procure two pipes, old ones preferred, but two clay pipes or a clay pipe and a corncob treated with a little paint or lacquer will do.

Fill the bowls of the pipes with glass wool. Amber colored glass wool is obtainable under the name of "Angel Hair" at stores selling Christmas tree decorations.

If boxes can be purchased or made to hold the pipes separately, so much the better.

Saturate the bowl of one pipe with concentrated hydrochloric acid, and the other with concentrated ammonium hydroxide. Keep them apart until ready to use. When placed next to each other, white smoke will appear and continue for some time.

Patter: "Whenever I have occasion to demonstrate some of the wonders of chemistry before an audience, I am reminded of my old chemistry teacher, Professor Adams. His habit when working in the laboratory was to smoke one or another of his pipes, of which he had a great number. The old man has long since passed to his reward, but I was fortunate in securing two of his pipes, which I prize highly as souvenirs. (Dislay the two pipes keeping them well apart from each other.) Curiously, whenever these pipes are placed near each other (place the second pipe near the first) the spirit of the dear departed Professor hovers over them and they smoke."

Smoke In The Bell Jar

A variation of the well known "Yogi Smoke Trick" which if properly presented can cause much amazement.

A large glass is swabbed inside with concentrated ammonium hydroxide and placed upright upon a glass plate.

A second glass plate has a few drops of concentrated hydrochloric acid in the middle.

Place three wine glasses on the table in a triangle and lay the plate on which the acid has been placed on top of them. Cover the glass containing the ammonium hydroxide with a hankerchief, lift it quickly and place it on the glass plate.

At a distance produce smoke by means of a cigarette, pipe, etc., and pretend to pass the smoke into the covered glass, which everyone can see is closed tightly and covered. When the cloth is removed, the glass is full of smoke.

The Atomic Bomb

(U. F. Grant)

The performer picks up an "atomic bomb" from the table—his latest discovery, so he says. A boy is asked forward to hold the bomb and is then asked to drop it into a can on the table so that the bomb will not explode. The performer then ploces a cover on the can to protect the bomb.

The boy is then instructed to hold the can over his head. As he picks up the can there is an explosion. Usually, the boy will drop the can from surprise. The performer then removes the cover and smoke pours out of the can.

For the bomb you can use one of the small red rubber balls sold for children. The can used should be one that has the bottom set up a ways. To the bottom of the can on the outside solder a bingo device. Put a cap in the device, set it, and then place the can on the table. When the can is lifted the bingo device will go off with a loud bang.

On the inside of the can on the bottom, place a few drops of ammonia. The cover of the can should be laying to one side and in this, place a few drops of sulfuric acid. When the cover is placed on the can and the fumes from the two chemicals units, and the can will be filled with smoke.

The Magic Pipe

(U. F. Grant)

The performer shows a miniature pipe and explains,"—it was once owned by a Magic Dwarf. Even though the dwarf has passed on he still likes to smoke the pipe." The performer places the pipe in a glass tumbler and covers the tumbler with a silk saying, "We will come back to that later and perhaps we may be able to catch the dwarf smoking the pipe."

He then goes on with another trick and a few moments later remembers the pipe, saying, "Let's get the pipe and try to get the Magic Dwarf back for a smoke." When the glass is uncovered, it is seen to be filled with smoke. The performer blows smoke from the tumbler saying, "Oh, I'm afraid that we are a little too late. The dwarf has already been here."

Use a miniature pipe obtainable from any pipe store. In the pipe goes the hydrochloric acid and in the bottom of the glass tumbler place the concentrated ammonium hydroxide. The rest is showmanship.

The Pirate Knife

(U. F. Grant)

This is a very clever item to use along with some other trick, such as the head or hand chopper. You explain to the boy assisting you that he must be very brave and must pass an old pirate test of bravery.

Taking the boy's hand you tell him that you will carve a cross brand on the back of his hand for the test. As you draw the blade of a "pirate knife" across the back of his hand in cross fashion, blood is seen to run from the apparent cuts. However, when you fan his hand, the blood vanishes and the cuts disappear, leaving his hand unharmed.

The knife used should be styled after a pirate knife. It can be made of wood or metal, but should not be sharp. Along the side of the knife runs a rubber tube with a bulb at the handle end. A red liquid goes in this bulb. As the knife is drawn over the boy's hand, the bulb is squeezed and the red liquid will run off of the blade and into the back of his hand.

To make this red liquid, which will disappear after a few moments, you will need some phenolphthalein in powder form. This can be obtained from any drug store. The powder is dissolved in alcohol and a little household ammonia is added to give it a red color. The liquid should be kept in a capped bottle just prior to the show. When the liquid is exposed to air (or fanning) the ammonia fumes evaporate, and the red liquid reverts to its former colorless state creating a most unusual effect.

The Spirit Glass

You show a piece of glass which is seen to be unprepared upon close examination by any member of the audience. Slowly as you hold the glass, a well-defined picture will appear on the surface. In a moment the picture will disappear even more mysteriously than it appeared.

To accomplish this miracle, write on a piece of glass about five by six inches in length with hydrofluoric acid. Do this by using a new steel pen and the acid as ink for drawing the picture desired on the glass. Allow the acid to remain on the glass for eight to ten minutes, then wash it off with distilled water and dry the glass by rubbing with a soft clean cloth.

Close examination will reveal nothing, but the picture will become instantly visible when breathed upon.

To determine the proper depth for the etching so the picture will be invisible when dry and yet clearly visible when breathed upon, experiment with several glasses letting the acid remain on each a different length of time.

Suggestions for this effect might be the use in card revelation or in mental and/or spiritualistic magic.

Magical Rainbow

Water from a pitcher is poured into a glass; and as you continue to pour into the glass, a series of mysterious transformations are seen to take place. As soon as you pour the water it turns green, then blue, next violet, slowly it turns to purple and finally in complete exhaustion it remains red.

The requirements for this effect are a chemical compound easily prepared at home without expensive or difficult equipment. Mix one part, by weight, powder manganese dioxide with three parts, by weight, potassium nitrate. Place in a sand crucible and bring to a red heat in a stove. Do not cover the crucible. In this way potassium oxide will be formed.

Allow to cool and secretly introduce a few grains into a glass tumbler. Pour water into the glass from a pitcher and the above mentioned series of color changes take place.

The Smoking Hands

Devoid of all "hook-ups" as smoke reservoirs, the Smoking Hands is an effect that is a very entertaining one as well as mysterious. The hands are cupped together and the thumbs are placed in the mouth as if they were a pipe. Slowly smoke appears from the hands as if one were using invisible tobacco lit with an invisible match.

With each puff (deflation of the cupped hands) smoke issues forth. The hands can be unclasped, showing them to be empty at any time.

The secret of the trick is in its prior preparation. On the palm of one, place some ammonia solution. On the other a drop or two of muriatic (hydrochloric) acid. To be sure of no burns, your hands must be free from all cuts or sores. Allow them to dry and perform the trick immediately. After the performance wash the hands in a solution of baking soda and rub in a small amount of a softening type hand lotion.

Gravatt's Mental Mystery

(Glenn G. Gravatt)

A number of blank-faced playing cards (obtainable from any magic dealer) are shown to have various simple letters and designs drawn thereon with heavy black ink. These are handed to a spectator. The magician retains a like number of cards with designs which are duplicates of the first packet.

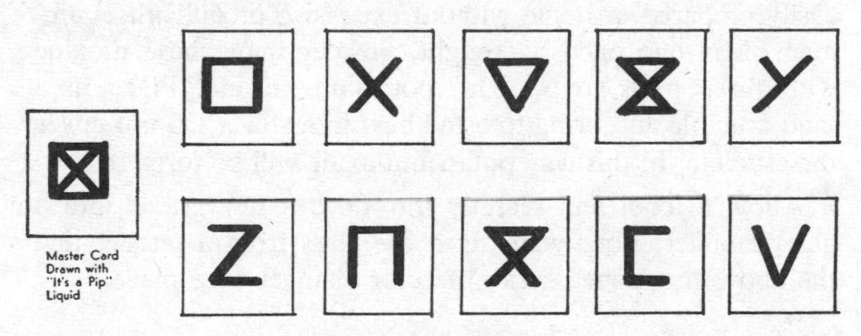
Master Card Drawn with "It's a Pip" Liquid

Performer selects one card and discards the remainder. He holds it in front of him and while he gazes at it and concentrates on it, spectator is asked to look over his cards and try to catch the performer's mental vibrations. Spectator tries to guess which card the mentalist holds and finally selects one from his packet which his strongest hunch tells him may be the duplicate of the one the performer holds.

The mentalist now emphasizes that had the spectator chosen his card first, some trickery might be suspected, but that he, the mentalist, chose a card, and after this was done, the spectator had free choice of any card held in his own hands. Performer calls attention to the fairness of the procedure. He says he hopes he is able, by sheer mental telepathy and concentration, to broadcast his thoughts to the assisting spectator.

Performer now shows his card. It is seen to be a duplicate of the other, thus proving the spectator's hunch was correct. The whole routine is simple, direct and straightforward.

Method: Magician's packet contains duplicates of spectator's packet, but with the addition of one "Master" card, on which is drawn a design with "It's a Pip" liquid, again obtainable from dealers. Magician holds his card (the Master card) back towards spectators, using both hands to hold the card. His two thumbs rest on face of card, fingers behind.

As soon as the spectator has decided on one card, selected it, and thrown it face up on the table, the mentalist uses either or both thumbs as the case may require, to erase one or more lines from the master design, so that his own card will duplicate the spectator's.

The time required to do this is very slight, and is covered by the mentalist's patter, reviewing the fact that he, the mentalist, chose his card before anyone, even the spectator, knew which one the spectator would select; that if the spectator has successfully received his thoughts, he will choose a duplicate of that held by the preformer; that if it turns out that they are not duplicate, it will be his, the performer's fault, for his in ability to send his thought waves out strongly enough.

Performer then throws his card face up on the table, and it is seen to match.

Ink and Water Columns

(Dr. W. M. Endlich)

This effect is founded on an old magical principle, but has been neglected during the past years. It is somewhat like the "Sympathetic Fluids," but in place of the red, white and blue liquids ink and water are used.

The performer has six glasses on a tray, with a glass cylinder about eight inches high and two inches in diameter, also a cardboard or fiber tube that fits over the glass cylinder.

The cylinder is covered with the tube (shown as empty),

then the performer turns to the glasses. Three contain ink, the other three water. He pours in ink and water alternately, emptying the glasses into the covered glass cylinder.

The audience would suppose that the result would be an ink-like liquid, but when the cardboard cylinder is lifted, he shows that the ink and water have formed in layers one on top of the other. This is very surprising.

Both the ink and the water are chemical compounds. To make the ink, take a saturated solution of tannic acid (spoonful to a glass of water). Now put a spoonful of this solution in each wine glass used for ink. To each of these wine glasses add two or three drops of tincture of iron (ferric chloride) and fill the glasses with water. This looks like ink:

For the three glasses of water, use a saturated solution of oxalic acid, half a glassful for each wine glass, and add water.

Inside the fiber tube is a celluloid cylinder which fits easily over the glass cylinder. This cylinder is covered on the outside with three bands of black paper (Mysto Tape) or black celluloid to represent ink. The bands may be painted on, in black if desired.

When working the effect, show the cardboard tube to be empty, cover the glass cylinder and pour in the liquids alternately. They change to a colorless liquid. When tube is lifted, the glass tube appears to have ink and water in layers, as the celluloid fake is left on the glass cylinder.

Replace the tube and stir the liquid with the wand—this is merely a byplay. In lifting the tube for the second time, carry the celluloid fake along with it, which will appear empty and can be shown as such. The glass cylinder will be filled with what appears to be water.

In handling these chemicals, especially the oxalic acid, it is highly important to remember that they are poisonous and should be used with great care. Never use the glasses for anything but this trick in order to avoid unnecessary accidents.

Sands of Enchantment

Take ordinary clear white sand which has been washed in water and colored by dyeing in any of several of the fast dyes on the market. Prepare a small quantity of each one (separate colors) by heating in a pan with a small piece of wax, stirring constantly. You will find the best way to determine the amount of wax needed by experimentation.

When properly prepared these sands will compress together and will also separate to flow as ordinary unprepared sand will. Place each of these colored sands in a separate open shallow dish.

Obtain a deep bowl of clear glass and fill with water. Dissolve in the water some tannic acid. Secrete in the palm of your hand a piece of blotter saturated with a solution of potassium ferrocyanide. Swish your hand around in the bowl of water and as the blotter wet with the ferrocyanide comes in contact with the tannic acid it turns black.

From each dish (separately) pick up a handful of colored sand and gently squeeze into a compressed ball. (This is unknown to the audience.) Place same into the bowl with already-dark liquid present. Repeat with other sands and then gently stir the liquid with your wand while telling your favorite story.

As your assistant brings onto the stage several clean dishes, reach into the dark water and remove one of the compressed balls of sand. As you hold it over the dish, gently allow it to flow from your hand in its original form. Repeat this with the others and as you remove the last, steal from the table a second piece of paper saturated with oxalic acid solution. Stir the liquid with your hand and it becomes clear "water" once again.

The effect to the audience leaves them in complete bewilderment, for the sands are separate in colors and completely dry and they can see through the bowl to see nothing (for there is nothing to see).

The Trick Without A Name

(Sid Fleischman)

In the effect, a spectator prints his name with red pencil across a page in a copy of the "Reader's Digest." The page is ripped out and the spectator retains the magazine. The performer sets fire to the page. As the flames devour it, the name in red can be seen to the last. A moment later the spectator is instructed to open the magazine, and he not only finds no trace of the torn leaf, but discovers the original page still in the book with the name in red still lettered across it

Red carbon paper and rubber cement make this effect work.

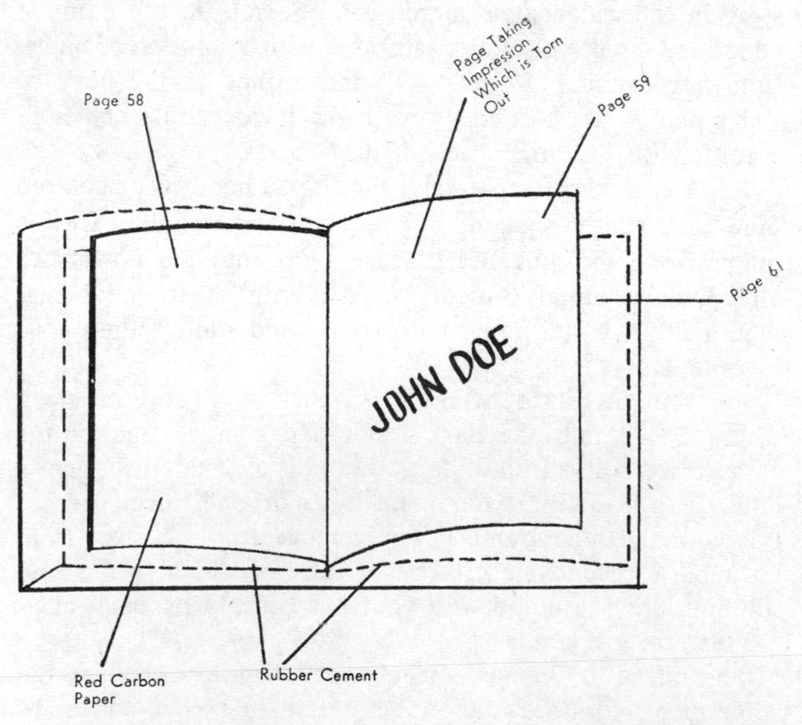

Open the magazine to a page near the center and run a thin stream of rubber cement along the three outside margins of the left-hand page—we'll call it page 58. Turn to page 61 (a right-hand page) and run rubber cement along its outside margins. As you know, two dry surfaces of rubber cement will adhere to each other. Page 59-60 lies between the two gimmicked pages and keeps them separated.

There is one more preparation to be made on page 58. At any large stationary store you will be able to buy red carbon paper. Cut a piece large enough to cover the body of the page and attach it there, business side up.

Turn back to page 54-55. Crease the binding so that the book will break open at that spot when casually opened.

When at a stationary store, pick up a red lead pencil, the harder the lead, the better.

The trick begins when you ask someone to say "stop" as you let the pages of a "Reader's Digest" flip by. Regardless of when the word is called, you manage to stop at the break-pages 54-55.

Handing a cooperative looking spectator the red pencil, direct him to print his first name across page 55. Call attention to the number. Ask that it be remembered.

With the name printed (rather than written—the average person bears down more when printing), take up the book. Close it up for a moment with your finger holding the place, deliver a line of patter, and in reopening the book you will find it surprisingly simple to let two leafs pass to the left. Immediately you tear out page 59—which bear the carbon image of the spectator's name. With the leaf out, close the book and give it back to the spectator. Now that page 59 is gone, the two surfaces of rubber cement (on page 58-61) come together and seal up the carbon on page 58!

Nothing remains but to burn the ersatz signature, have the book reopened and discover the original still in place. The magazine can be freely examined, The originator of this effect (Mr. Fleischman) has performed this effect many times and no one has yet discovered that two leaves were joined together.

How to Treat Rope Ends

Trim the ends of the rope diagonally (at an angle) so they slope down to a sharp point.

Dip the ends of the rope one at a time into "rope cement" which your magic dealer can supply. Revolve them so they will coat evenly all around.

Hang up to dry in such a way that the ends will not touch anything.

Allow to dry from twenty-five to thirty minutes, longer on a damp or rainy day. It is important to allow them to dry completely.

Do not touch the ends of the rope before or after preparing the ends as outlined in the foregoing.

To use the ropes prepared in this way for a cut and restored rope routine, follow the directions described next. Bring the two "treated" ends together so they slightly overlap, about three-eighths of an inch is best. Now taking the joint between the fingers and thumb, roll it gently at first. Increase pressure slightly until the joint is pressed down to the same size or thickness of the rest of the rope. After a few trials you will get the idea and be able to make a neat joint that will be invisible at a slight distance. As rope is in constant motion when you are performing the trick, it will be hard to find the location of the joint without looking too closely, which would be suspicious.

To locate the joint at any time as required in several of the rope tricks on the market, without sufficient close inspection, just pull the rope through your fingers in a natural manner and you can feel the difference in the surface of the rope.

Frozen Smoke

(Earl C. Leaming)

Procure a plain unfinished cardboard mailing tube about one or one and one-half inches in diameter and from ten to twelve inches long.

Take a packet of ordinary book matches, the kind having yellow stems.

In a small dish place about half a teaspoonful of flowers of sulphur. Heat gently until the sulphur is melted. Using a toothpick, take up a drop of molten sulphur and quickly apply it to one of the paper matches, placing the sulphur just below the head of the match and upon the yellow stem. Apply with a quick downward stroke so as to deposit the sulphur in a thin layer about one-third of the length of the match stick. Treat several matches in this manner, or one match only in several packs. If the matches are to be passed for inspection by the audience it is necessary to have but one prepared match in the book. If, however, you wish to use a book as stock, provide a second similar unprepared book and make a switch or palm a prepared match for your own use.

Provide a small vial in which a teaspoonful of stronger ammonia water is placed. The vial should be well-capped until using.

Presentation: Just before you are ready to perform this effect, pour the ammonia down inside the mailing tube, taking care to rotate the tube and pour so as to have no visible wetness on the outside. Avoid getting ammonia on your fingers or in your nose.

Holding the tube vertically in one hand, strike a prepared (sulphur) match and hold it under the lower end of the tube, yet down far enough to allow the spectators to see that the match itself does not smoke. A cloud of dense white smoke will appear at the top of the tube. Take the match away and the smoke will stop. Apply the match and get more smoke.

After all the ammonia has been used up, both tube and unprepared book of matches can be passed for inspection.

Patter: "You will no doubt remember that upon one occasion the Baron Munchausen was invited to spend a few days at hunting with a Russian Nobleman friend. The castle-like hunting lodge of this Nobleman was in faraway Siberia and the temperatures extremely low. In fact, upon the second day the notes froze in the huntsman's horn. Later, hung by the fireplace, the horn thawed out and played the notes just as the huntsman had put them in. My own experience while not quite so strange was curious enough. While travelling through Saskatachewan during a cold spell last winter, we were told that the temperature had dropped to seventy-five below zero. I wanted to smoke very much, but there being only two cars on the train and no smoker and an old maid in the front seat objected so strongly that I looked about for some means of overcoming my predicament.

I noticed that there was a broken window into which a rag had been stuffed and then I found this old piece of cardboard tube.

Quickly I removed the rag from the broken windowpane and adjusted the tube into the hole. Lighting a cigar I proceeded to enjoy my smoke, puffing out through the tube. My enjoyment was shortlived, however, for to my surprise the smoke froze in the tube.

To prove it to you, I will light a match and thaw out what remains."

Magnet-Tizo
(Burling Hull)

This remarkable compound has many peculiar qualities that make possible the most mysterious magical effects possible.

Test Experiment: First be sure that your fingers are thoroughly clean. If you have just washed them, then dry them most thoroughly before attempting the experiment.

Scrape a thin wafer from the Magnet-Tizo. Press between the fingers and thumb tips, working it so it warms and softens and coats the surface, or you may melt Magnet-Tizo and then apply it.

Next wipe off all the excess from the fingers with a clean handkerchief, until you can no longer see any remaining. There will still be a thin invisible coating left in the pores and over the skin surface.

Thus prepared, show your hands around to those near you. Draw your fingers smoothly and lightly over the hands of a spectator showing that there is not the slightest "cling" or stickiness to your finger tips.

Performance: Take a borrowed card, pencil, ruler, coin, a light magical wand (the paper wands with wooden tips used in the Vanishing Wand trick work well), a polished ball, pen—in fact any article not too heavy. Have an egg with the contents blown out (explain this is because of the danger of having an accident) and pass for examination.

Take article between fingers and thumb of either hand, press tightly between them (and sort of slide the fingers over the object) and then the article will cling most mysteriously to the finger tips. Transfer it from thumb to fingers and back, from hand to hand, passing hands right before and close to spectator's eyes, turning hands on all sides, showing freely.

State you will "demagnetize" the object. Make a "snap" with the fingers of other hand, or pass the empty hand over hand holding object, and at the same instant move the second finger forward and the first finger backward slightly. This will have the effect of breaking the light adhesion of the article, which will drop.

With a little experimentation a number of most weird,

mysterious and uncanny effects can be created by clever showmanship.

Silk handkerchiefs work beautifully in the same way.

As a suggestion, have the spectator take hold of the article asking him to feel the magnetism.

A Flash Paper and Match Stunt

(Ireland Year Book)

Make a small ball of flash paper and lay it on an ashtray. Now strike a match, light a cigarette and blow the match out. Then touch the same match to the ball of flash paper. The flash paper will burst into flame and relight the match. This is a very weird stunt and will cause no end of surprise. The match is actually blown out, and you need not move too fast in putting in against the ball of flash paper. Try this out and you will first astonish yourself, then get a lot of fun trying it out on others.

Producing Two Flashes of Flame

Here is how to produce two flashes of flame when lighting a cigarette. Several features of this are that there is no danger of burning the fingers; and after you cause the flashes, everything can be handled naturally and no trace is left.

Requirements are a square of flash paper, a regular pack of cigarettes, a bit of magician's wax and a box of safety matches.

Method: Tear off two small pieces of flash paper. One piece is about one-fourth inch square and the other about one-eighth by one-half inch. The latter is narrower than it is long. Roll the square piece into a small pellet. Take a cigarette and with a needle dig out a bit of tobacco from one end of the cigarette. Push the pellet into this cavity. Place the cigarette back into the package. The pellet is rolled tiny, but care must be taken so it will make a pretty noticeable flash when set off.

The longer strip of paper is wrapped around one of the safety matches at the unburnable end at its tip. It is wrapped tightly and several times due to its length. A tiny bit of magician's wax is now pressed on the last corner of the flash paper to keep it from falling off.

This prepared match is placed back in the box with the remaining matches. You are all set to begin. Place the matches and cigarettes in your pocket.

Performance: Take the cigarette package from your pocket and remove the prepared cigarette placing it between your lips with the bit of flash paper at the outer end. Turn your right side to the audience as you reach for the box of safety matches. This will insure you against some brilliant eye noticing the pellet of flash paper.

Holding the box of safety matches in both hands, open it up and remove the match, unprepared and in position to strike it on the side of the box in the usual way of lighting a cigarette. The right fingers shield the flash paper from the view of the audience. Strike the match on the box and quickly light the cigarette. This causes a flash. Shake the match to douse it immediately after the cigarette is well lit. Puff several times and exhale the smoke to show the cigarette to be lit.

The match has cooled enough by now to grip the head of it between the right first and second fingers near their tips. The right hand holding the match clipped thus comes up in front of the cigarette momentarily and touches the prepared tip of the match to the lit end of the cigarette. This causes a flash which melts the wax.

Toss the match aside and use the lit cigarette in your mouth as the opening for some cigarette work.

A Self-Lighting Candle
(Ireland Year Book)

Take a candle about four inches long. Drill a hole in the candle close to and parallel with the wick. This can best be done by heating a stiff wire to a warmth that readily melts the wax, and pushing it straight up alongside the wick. When it will not go any further, pull out, reheat, and continue to push.

Make the hole a little larger than the glass tube supplied with the Ronson cigarette lighter fluid. Attach a rubber bulb on one end of the tube, so as to make it like a medicine dropper. Fill the tube with alcohol and slip into the hole in the candle. Candle wick has a small piece of potassium metal on the base of the wick. Now by holding the candle in the outstretched hand and slowly squeezing the bulb, a drop or so of the alcohol touches the potassium and as the alcohol is also fuel it lights the potassium metal which in turn lights the candle. Candle can be lifted off the glass tube at once with the other hand, if desired.

Secret Message

Obtain a sheet of unglazed writing paper, a solution of phenolphthalein in alcohol, and a sodium silicate solution. With a clean pen, write your intended message using the phenolphthalein as ink. On the opposite side of the paper, paint with the sodium silicate solution. Set aside to dry.

Exhibit both sides of paper to the audience, having same initialed. Fold paper in half, letting spectator hold until ready for revelation. For safe keeping place in book (for pressure) with edge sticking out. When spectator removes paper, message in red lines will be present.

The secret of this effect is that in folding the paper, the invisible phenolphthalein picture must come in contact with the sodium silicate.

Balloon "Skullduggery"

The effect in brief is to have a youngster come up on the stage to assist you in doing magic. Hand him a balloon to blow up; he does so, when suddenly it bursts as you take it from him. Handing him a second balloon, he is unable to blow it up as hard as he might try. The magician takes the balloon from his volunteer assistant, ties a knot in it and places it upon his table. All of sudden it is seen to blow itself up of its own accord.

The secret is simple. First you must have secreted in your hand a small pin, with which to puncture the first balloon as you reach for it. The second balloon is prepared by already having a tear in it. In this way it will be impossible for anyone to blow it up. The final bit of preparation is to have ready a prepared balloon for self inflation. The mehods for doing this are discussed elsewhere in this book.

The only necessary moves in this effect are to be able to switch the chemically prepared balloon for the second balloon (one with a cut in it). Perhaps the easiest way is to have the prepared balloon already on your table. Take the balloon from the volunteer assistant and make a motion to place it on the table. In reality you actually start the chemical reaction so the prepared balloon will inflate itself. As it does this, calmly secrete the balloon in your hand in your pocket.

Try this effect and the laughs you will get at the expense of the volunteer assistant will more than make up for the little amount of preparation needed.

Floating Metal Disc

A puzzling effect which requires but one subtle move. You must be able to switch a piece of metal the size of a quarter for another one. As magicians are adept in the art of sleight of hand, this should not be difficult.

Cut from a sheet of aluminum a round disc the shape of a quarter. Do a similar thing with a sheet of zinc metal.

Hand the spectator the zinc disc. Do not allow the fact known that a duplicate one of aluminum is in your possession. Ask the spectator to see if the metal disc will float on water. Offer him a glass full of water for trial.

It will be seen that the disc will sink to the bottom. Take this disc from him and wipe it dry, switching it for the aluminum disc; the magician attempts to make it float and he succeeds.

A good effect possible to tie up with the power of concentration theme, or the shift of atoms and molecules to cause strange things to happen.

Pearls of Buddha

Showing a green glass object, the shape of a tear, the magician explains that it is the pearl from a statue of Buddha found in the ancient forests of the mystic Orient. He continues to explain that if ever stolen, the curse of protection will follow these pearls, for if one should ever chip or be dropped and broken ever so slightly, they would shatter in countless pieces of useless glass.

By way of illustration, the magician exhibits one of these pearls, calmly striking the table with it. Nothing happens. However, he then asks a volunteer spectator to break off the fine tip from the end of the pearl and as he does, it breaks into thousands of fragments.

To make these pearls, you must melt some common green glass. While red hot, drop into cold water. The mass will assume a tear-like form. The spherical portion will bear very rough treatment, but the instant the smallest particle of the tail (tip) is broken off, the whole tear (pearl) breaks with a sharp noise into countless fragments.

Magic Metal

As an interesting addition, the author has included in this book the formulas for two metal alloys that when prepared will melt in boiling water and may be held in the mount without danger of burning. These are the metals that resemble lead and are usually found as such in the repertoire of fire-eaters.

It is not advisable that the novice try this effect as the danger of swallowing the melted metals offers a precarious problem.

I

Melt in a suitable dish bismuth metal, by weight, five ounces; lead metal, by weight, three ounces; tin metal, by weight, two ounces.

II

Melt in a suitable dish bismuth metal, by weight, eight ounces; lead metal, by weight, four ounces; tin metal, by weight two ounces; cadmium metal, by weight, two ounces.

Magic Dye

Prepare a strong solution of indigo sulfate with an equal amount of potassium carbonate. Dissolve these in distilled water.

A piece of cloth previously dyed various colors will exhibit strange results when immersed in this solution and subsequently removed, as follows: white cloth becomes blue, red cloth becomes violet, yellow cloth becomes green, any vegetable dyed cloth will turn red.

Ideas for Routines with Chemical Magic

This short discussion was included because of a statement made by a well-known magician in regard to chemical magic. "Chemical magic is good because it's different. It lacks in one point, mainly the convincing fact that this type of magic has long been associated with a childish type of legerdemain. It is, however, without a doubt a very simple type of magic which will seldom fool an audience, but with clever routining the simplest chemical color change can be created into a very mystifying type of magical illusion."

With these thoughts in mind, further thinking developed into a few basic ideas which ought to bring forth that chemical magic can not only be a simple type of magic, but also it is capable for chemical magic to carve a niche amongst other magical fields of legerdemain.

Chemical Substitution

Filling a large glass with water (use solution discussed in effect "Clock Reaction-White") and covering it with a cylinder, the magician places a glass plate over the top of the cylinder and on top of this he stands an empty milk bottle. This he also covers with a cylinder and pours milk into it filling up the milk bottle. The milk bottle has previously been prepared by coating with a strong solution of hydrochloric acid. The milk is a mixture of ferrous ammonium sulfate and strontium chloride dissolved in distilled water.

What happens is obvious, the water in the bottom glass turns to milk in a matter of seconds and the milk in the bottle turns to water. When uncovered, the milk and water have changed places.

Ink — ???

A simple effect the author has had a lot of fun with is to ask a person to fill his pen up with ink and when he starts to write it is water. To prepare for this effect obtain the type of ink well that has a little cup in the top which holds enough ink after inverting the bottle. In this cup place a strong solution of sodium thiosulfate. The ink is the iodine-potassium iodide solution. When the ink well is inverted, the thiosulfate reacts with the ink to turn it colorless (water).

Another routine similar to the above is to have the thiosulfate solution in the pan and when opening up the valve of the pen to remove the air prior to filling, the thiosulfate solution is ejected into the ink where the transformation takes place.

Color Changes

Pouring water into several glasses with different colors resulting, one always hears the word chemicals from the audience. Why not dispense with the glasses and use white cloth flowers prepared as follows: the leaves and stems are treated with malachite green, the flowers are treated with phenolphthalein and in a watering pot dissolve some ammonium hydroxide solution. Watering the flowers with this solution causes red flowers with green leaves and stems to appear. Leaving a few of the flowers unprepared, white flowers will remain midst the red ones. This causes an additional bit of mystery.

Chapter **XIV**

Chemical Tips For Better Tricks

A successful magic act is often accomplished by several important factors. Amongst them is the knowledge of the use of several "magical" preparations. Known only to the magician, they enable him to seemingly accomplish "miracles."

Such preparations are the use of daub, roughing fluid, slicking paste, pip paint, etc. Included in these pages are a few of the more important types of preparations. Use them to your advantage, for with their use you will be able to perform "miracles." These "miracles" of a fact come from secrets, so guard these secrets with care. They are the source of how to perform "real magic."

Daub

Powdered soapstone .... 39 parts by weight
Oleic acid .... 9.25 parts by weight

To prepare daub by the use of the above formula, incorporate well into the oleic acid with dye (color) of choice by vigorous grinding. Than add the powdered soapstone and continue to rub well until a uniform mass is prepared. Store in dark-colored ointment jars with a tight cover.

Metallic Daub

Powdered soapstone	39 parts by weight
Hydrous lanolin	6.5 parts by weight
Yellow beeswax	5.25 parts by weight
Stearic acid	3.25 parts by weight
Mineral oil	7.5 parts by weight

The above mixture should be melted over a water bath, removed from the heat and then add 0.8 parts by weight of powdered aluminum metal. Stir until completely blended. Store in dark-colored ointment jar with a tight cover. Cool before using.

Daub Substitutes

In lieu of standard daub preparations when they are not available, there are several effective and well worth considering substitutes that may be used.

The silver water color used by artists is one often used with considerable merit. The use of leg-make-up, rouge, lipstick, even mud scraped from the bottom of one's shoe might well fit in with an imaginative mind. To effect a substitute, one requires an imagination. Remember, you alone know what you are doing, thus the audience, not being aware of any planned deception, might continue to be fooled by a magician capable of showmanship and nerve.

Milk Pitcher Magic

The use of milk will often stain milk pitchers. These stains are difficult to remove and offer a problem. An answer to this problem has been found in the use of a solution of trisodium phosphate. Pour a little into the pitcher and allow to stand. Rinse out with clear water and dry.

That Egg "Again"

The use of a "feke" in the egg and fan trick is necessary. Following is a method by which the magician might make his own and have several spares on hand.

Preferably using a "blown" duck's egg, immerse it in a solution of glacial acetic acid and water, 1 to 6, until the outer shell is dissolved, leaving just the inner membrane.

Wash well in warm water and dry. This results in a "feke" which will pass for an egg at a very short distance.

Beverage Substitutes

A good substitute for beer is made by filling a glass half full with freshly made coffee and fizzing it up with a fast stream of carbonated water.

Burling "Volta" Hull of Deland, Florida has on the market a product known as "Beer Powder;" use of this as per directions included will result in a glass of foaming beer difficult to distinguish from the original.

To create coffee, dissolve a sufficient amount of powdered extract of licorice in warm water.

As a milk substitute, the product known as "OOM" supplied by your magic dealer is probably one of the best available.

In the event you are out of the above preparation, a teaspoonful of dried milk will usually work when dissolved in a glass of water.

Flesh Paint

A good substitute for flesh paint is the make-up sold in lieu of hosiery for ladies. In order to make it permanent, mix it with a small quantity of flexible collodion or with a small amount of varnish.

Wax Substitutes

Modelling clay of the soft non-drying variety usually sold in toy shops has been found to be an adequate substitute for magician's wax.

A small piece of well-chewed gum will often work in an emergency. Magnet-tizo, a product of Burling "Volta" Hull, is one of the superior products used for wax in magic. Furnished with this is a manuscript outlined with full directions for use of this product.

To those who wish to make their own wax, here is a formula. Melt in a water bath 8 grams chalky Carnauba wax, 5 grams dry paraffin wax at a temperature not higher than 160 degrees F. Heat 60 cc. naphtha to 120 degrees F. in a water bath by placing it in boiling water (all flames must be extinguished as naphtha is highly flammable). Mix the melted waxes with the hot naphtha. Pour into suitable containers and allow to harden.

For Slicking Cards

Two of the most desirable preparations on the market for use in preparing cards so they will be slick and will fan easily are ordinary floor polish, rubbed down to a high gloss and zinc stearate baby powder, often sold under the name of fanning powder.

Roughing Fluid

Dissolve in a mixture of 8 ounces of grain alcohol and 1 ounce of liquified phenol the following: one-half ounce, by weight, gum sandarac; one-half ounce, by weight, gum mastic. Agitate the bottle to hasten the dissolving. Allow to stand several hours and then filter to eliminate any sedimentation.

To apply roughing fluid, moisten a piece of cotton and apply to the desired surfaces by rubbing across the entire area with a back and forth stroke. Set the card aside to dry.

Pip Paint

Prepare a mixture of 3 grams carbon black in 30 cc. distilled water. Add a little of this mixture at a time into a second bottle containing 90 cc. liquid latex. Shake the bottle thoroughly after each addition of the color mixture.

Dampen the surface on which the pip paint is to be applied with cotton wet with distilled water. Apply the pip paint using a fine sable brush.

Pip paint which thickens due to evaporation may be thinned with distilled water to which has been added a few drops of ammonia.

Silk Cleansing Compounds

Grease spots when sprinkled with talcum powder and rubbed off by use of a soft flannel cloth will rid the stain from the material.

To dry clean silks, first wash in benzine and then wash thoroughly with soap suds to which a small amount of ammonia was added. Rinse in cool clean water and dry thoroughly in the open air.

Another preparation for cleaning silks may be made by mixing as follows: one and one-half ounces of borax, one-half ounce soap, one-half pint alcohol, one-half pint distilled water, one-half ounce magnesium carbonate and the yolks of two eggs.

Apply the mixture to the stain and then wash in lukewarm water. Rinse in cold water and dry at room temperature. Silks may be ironed with a slightly warmed iron.

Cleaning Feather Flowers

Saturate them with benzine and allow them to dry in the open air. Do not inhale the vapors.

Exposing feather flowers to steam vapors will freshen them.

Removal of Stains from Cards

Spirit of camphor rubbed on spots on playing cards with a small piece of cotton has been proven effective in the removal of same.

Fire-Proofing

To render cloth materials fire-proof, impregnate them with a solution made by dissolving a teaspoonful each of boric acid and borax in four ounces of water. After soaking, allow to dry thoroughly before using.

A second method is to follow the same procedure outlined above except substitute the following: A saturated solution made of borax, ammonia alum and sodium chloride.

Stains from Metal

Egg stains will often appear on the nickel-plated apparatus used by magicians, (dove pan, welst rarebit pan, etc.) These stains can be removed by rubbing with common salt and then rinsing in lukewarm water.

Nickel-plated articles which have become dull can readily be restored by means of alcohol to which 2 percent sulfuric acid has been added. Apply the cleaner liberally and then wash off with clean water after a few seconds. Polish with pure alcohol and a dry cloth.

Dissolve 5 grams oxalic acid in a pint of hot water. Allow the equipment to soak in this solution. After the solution has cooled, remove the equipment and rinse in cold water. Dry thoroughly.

Cement for Glass

Mix 10 parts casein with 60 parts water glass solution. Apply the cement as quickly as possible and dry in the air.

Here is a cement worthy of consideration when you break a piece of equipment made of glass and highly valuable to you.

Diachylon

Dissolve in three hundred fifty cc. distilled water, one hundred grams, by weight, dried, coarse soap powder. The water must be hot. After stirring, strain the mixture. Dissolve sixty grams, by weight, lead acetate in two hundred fifty cc. hot distilled water. Add this second solution with constant stirring to the soap solution. Decant off the supernatant liquid and wash the precipitate with hot water. Do this several times.

Transfer the mass to a warm slab and knead thoroughly to free it from water. Roll in cylindrical forms and wrap with paper until ready for the second operation.

Melt the lead plaster made in the foregoing by gentle heat, using fifty grams, by weight. To this add forty-nine grams, by weight, olive oil. Mix this mixture thoroughly and allow to cool. Then add one gram, by weight, synthetic oil lavender. Stir until the diachylon begins to congeal and then pour into suitable containers.

Chapter **XV**

Chemical Jokes and Novelties

Exploding Matches

In preparing this mixture you are handling a powerful chemical liable to explode at the slightest touch. Handle with extreme care and make it in the smallest quantities you are capable of working with. Do not pour any of the mixture or waste products down the drain. REMEMBER—UNLESS YOU KNOW WHAT YOU ARE DOING AND HOW TO DO IT? DO NOT PREPARE THIS MIXTURE.

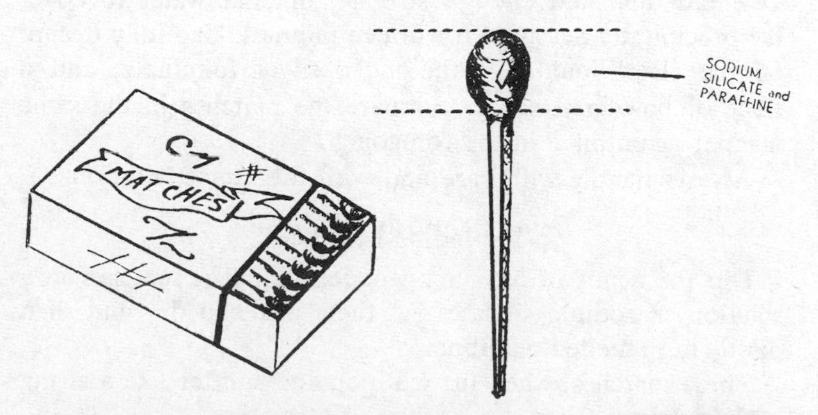

Mix two grams of iodine crystals with enough ammonia solution to just cover the crystals. Allow to stand for ten to fifteen minutes. Decant off the supernatant ammonia liquid.

Repeat once more with ammonia and then once using distilled water. Do not allow the precipitate to dry until you have disposed of it. (Caution: When wet, this mixture is safe to handle. When dry, observe the cautions outlined above.) Mix a little powdered acacia with the wet precipitate and while still wet place a *minute* quantity of it on a match just below the head. Allow to dry and you will have prepared exploding matches.

Exploding Matches

As in the preparation of the first exploding match, the preparation of silver fulminate is another dangerous operation. Exercise all the cautions outlined in the foregoing.

Dissolve seven and one-half grams of silver nitrate in thirty cc. water (distilled). Filter and add a solution of fifteen grams sodium hydroxide dissolved in thirty cc. distilled water. Add a little at a time until the black precipitate is completely formed. Filter and wash the precipitate with distilled water. Dry and then add enough strong ammonia water to cover the precipitate. Set aside for fifteen minutes. Carefully decant off the clear liquid. To the moist silver fulminate, add a trace of powdered acacia. Prepare the matches in the same manner as outlined in the foregoing.

Always handle with care and with the utmost caution.

Exploding Matches

Dip the heads of ordinary wooden or paper matches in a solution of sodium silicate. Set them aside to dry and then dip them in melted paraffine.

These matches, when lit, will pop and sputter like a string of firecrackers.

Diabolical Candles

Into paper tubes about one inch in diameter hold in a vertical position and sealed at one end with a wooden block, place a piece of string. Next pour in a mixture of melted paraffine in which has been mixed some melted sulphur. When fully hardened, remove the paper tube and trim off the string to form a wick of a candle.

Lighting one of these candles will produce the odor of rotten eggs.

Trick Matches

Dip the heads of matches into a solution of sodium silicate. Set aside to dry. When struck, they will sputter and will not light.

Sugar — ???

Coat lump sugar with flexible collodion. Set aside to dry. They will then be with a water-proof coating. When used in coffee or tea, they will float and will not dissolve, much to the consternation of the user.

Serpent Matches

Mix together 8 grams of powdered potassium bichromate, 4 grams of powdered potassium nitrate and 4 grams of pow-

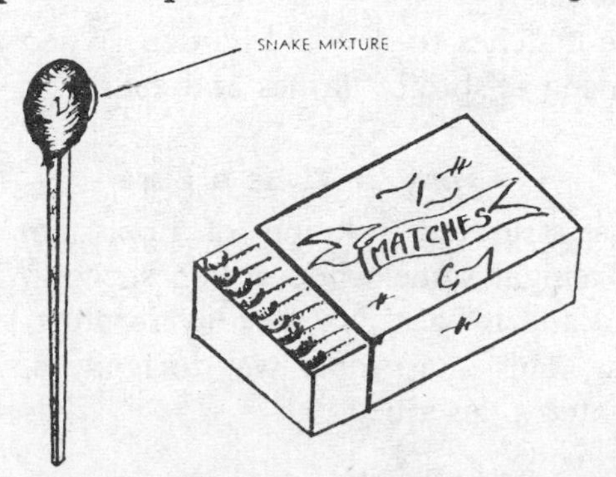

dered sugar. Moisten with just enough mucilage of acacia so that it will hold its shape when molded. Apply a little just below the head of the match. Set aside to dry.

Lighting the match in the usual manner will cause the appearance of a long "snake" like substance to appear. The size will depend upon the amount of the mixture used. The "snake" will appear to materialize from within the flame.

Vanishing Ink

A solution made by dissolving a small quantity of phenolphthalein in alcohol and added to ammonia water will appear as, and may be used as, red ink.

Writing with this "ink" and then exposing it to the air will cause it to vanish. Since the ammonia is highly volatile, it evaporates from the paper and the writing becomes colorless.

Sparkling Matches

Mix, but do not grind the following: 4 parts, by weight, powdered magnesium metal and 2 parts, by weight, powdered potassium nitrate. Melt some paraffine wax and dip the matches into same. While the paraffine on the match is still hot, roll the matches into the magnesium-potassium nitrate mixture. In this way the head of the match becomes coated with the powders.

Allow the matches to dry and harden. When lighted, they will burn giving off bright flashes and sparks.

How to Make a Fuse

Prepare a saturated solution of potassium nitrate and soak in it ovenight some white cotton string. When dry, ignite one end and it will burn steadily forming as it does an effective fuse. This is a proper way to ignite any of the colored fire mixtures described.

Magic Cigarettes

With a weak solution of sulfuric acid made by dissolving the acid in distilled water, write a secret invisible message

on each cigarette, using a clean pen. When the cigarettes are lit, as they burn the message is slowly revealed.

Diabolical Bombs

Although not recommended as the best way to get along with society, here is the method by which one can make a diabolical bomb which will undoubtedly have one of the most offensive odors to be created.

Obtain some glass tubing which you will seal on one end by placing it in a hot flame (i.e. bunsen burner). When the end is sealed and fairly soft, gently exhale through the mouth into the open end of the tube after removing from the flame. In this way a thin glass bubble will be formed.

Fill same with the solution of ammonium sulfide. Do not fill it completely full. Seal the other end of the tube with a sealing compound. The device is now ready for use by breaking the bubble and "running." However, as so stated in the beginning, it is not advisable to make these nor use them.

Chameleon Powders

Certain chemical powders when subjected to physical forces will change colors. A few of the more common changes are herewith illustrated.

Mercuric oxide (red) when heated will change colors as follows: First it becomes a bright red, then dull, changes to purple, black and finally upon cooling changes back to its original red-orange color.

Mix equal portions of white potassium iodide and white powdered lead nitrate. Place in a small cardboard box and after showing, ask the color. The response will be, "White." After shaking the box rather vigorously, hand it to the spectator to open and he will be surprised to see instead of a white powder, a yellow powder.

Living — ???

Certain substances when placed in contact with water will dart about on the surface as live insects might do.

Drop a minute particle of sodium metal to achieve this effect. It will dart back and forth with a crackling noise.

Minute bids of camphor (natural, not synthetic form) will also respond to a stimulus such as a push with a pencil, and then continue to move about over the surface of water.